Electrical Installations in Hazardous Locations

Second Edition

Peter J. Schram and Mark W. Earley

National Fire Protection Association
Quincy, Massachusetts

Product Manager: Jim L. Linville
Project Editor: Kimberly Cervantes
Copy Editor: Amy Lewis
Text Processing: Elizabeth Turner
Art Coordinator: Nancy Maria
Manufacturing Buyer: Ellen Glisker
Interior Design Coordinator: Cathy Ray
Illustrations: George Nichols, Todd Bowman
Original Photography: Richard Green
Index: Hagerty and Holloway
Composition: Modern Graphics
Cover Design: Groppi Design
Printer: R.R. Donnelley & Sons

One Batterymarch Park
Quincy, Massachusetts 02269

NFPA No.: HLH-97
ISBN: 0-87765-423-9
Library of Congress Card Catalog No.: 97-075554
Second Edition: 60124

Printed in the United States of America
10 9 8 7 6 5 4 3 2 1

Contents

Preface

Electrical Installations in Hazardous Locations is a single source of information designed to complement the many codes and standards produced by the many U.S. and international standards-making organizations concerned with the installation of electrical equipment in hazardous locations.

Intended to serve as a general reference — not the final word on matters of code interpretation — *Electrical Installations in Hazardous Locations* begins by thoroughly explaining exactly what hazardous locations are — and what they are not. From there it progresses, in a linear fashion, to more complex issues. Among the hundreds of important subjects it covers are the following:

- How to define and distinguish hazardous locations: classes, groups, and divisions
- Requirements for equipment protection systems
- Protecting against ignition from static electricity and lightning
- The distinctions between American and international requirements

Electrical Installations in Hazardous Locations was written with the specific intent of advising people who are unfamiliar with, or who only occasionally become involved in, the intricate details of electrical installations in hazardous locations. Until now there has been no single source for this general information. It is the authors', and the editor's, hope that this book will fill that void.

Acknowledgements

The authors wish to thank all of the many people who helped, either directly or otherwise, in preparing the material that appears in this book, especially our wives and families. Without their support this book would not have been possible.

Special thanks are due to Alonza W. Ballard, manager, Technical Standards and Compliance, Crouse-Hinds, Division of Cooper Industries, who revised Chapter 6, ''Requirements Outside the United States''; Dann Strube, a private electrical code consultant based in Lanesville, Indiana, who revised Chapter 3, ''Classification of Specific Processes/Occupancies''; Robert P. Benedetti and John M. Caloggero, senior fire protection engineers on the staff of NFPA, who revised Chapter 5, ''Static Electricity and Lightning Protection''; and Gary Davis of HEP–Killark for permission to reprint many of the illustrations.

About the Authors

Recently retired, Peter J. Schram was NFPA's chief electrical engineer from 1980 to 1989. Now a private consultant specializing in electrical safety, he serves as a special expert member of the National Electrical Code Committee — Panel 14, which is responsible for the development of Articles 500, 501, 502, 503, 504, 505, 510, 511, 513, 514, 515, and 516 of the *NEC®*. While employed by NFPA, Mr. Schram served as the staff liaison to the National Electrical Code Committee and as secretary of the *National Electrical Code Correlating Committee.* He was the editor of the 1984 and 1987 editions of the *National Electrical Code Handbook*. Prior to joining NFPA he was with Underwriters Laboratories Inc for 30 years, including 10 years as associate managing engineer of their Hazardous Locations Equipment Section.

Mark W. Earley, P.E., is assistant vice president of the Engineering Division at NFPA. He is secretary of the committee on the *National Electrical Code®,* the U.S. co-secretary of the correlating committees on Electrical Installation, an associate member of the Canadian Electrical Code Part I, and a member of the Electrical Council of Underwriters Laboratories, the Electrical and Electronics Standards Board of the American National Standards Institute, the Executive Committee of the U.S. National Committee on the International Electrotechnical Commission, and the Safety Council of Inchcape ETL Laboratories. Mr. Earley is also a member of the International Association of Electrical Inspectors, the Institute of Electrical and Electronics Engineers, and the Society of Fire Protection Engineers. He was the editor of the 1990, 1993, and 1996 editions of the *National Electrical Code Handbook.* Prior to joining NFPA he was with the Factory Mutual Research Corporation for eight years.

CHAPTER 1

Introduction

1-1 Why This Book Was Written

There are many publications relating to hazardous locations, including standards for the construction and performance of equipment, installation standards, and standards and manuals classifying certain types of hazardous locations. In addition, books and technical articles, some of them highly technical, have been published relating to specific aspects of hazardous locations. However, until the first edition of this book was published in December 1988, there had not been available a general book written with the specific intent of acquainting someone unfamiliar with the subject or who must only occasionally become involved with all of the various aspects of installations in hazardous locations.

This book was originally written with the intent of filling this perceived void in the literature. The authors hoped that it would be of value to those needing information on hazardous locations as they are defined in the United States. Information was also provided on requirements outside of the United States.

This second edition has the same intent as the first. It has been updated to represent current technology, and the references were modified where necessary to show the latest national and international standards information. The information has also been expanded and new material has been introduced.

This book is not intended to provide details on the requirements that appear in the many different standards published on the subject. By necessity, however, some of these requirements must be included or referenced. Wherever a document is referenced the reader is urged to check the latest edition of the document by obtaining a copy from the organization that is responsible for it or by checking with that organization to be sure that the document referenced in this book has not been changed, such as through the publication of a new edition. Most standards are updated frequently.

1-2 The National Fire Protection Association

The National Fire Protection Association (NFPA) is one of a number of organizations publishing standards relating to hazardous locations. Like many standards published by other U.S. organizations, those of the National Fire Protection Association are processed through the American National Standards Institute (ANSI) and are identified as American National Standards by the identifier ''ANSI/'' preceding the identifier of the publisher of the standard and the publisher's number for the standard. For example, the National Fire Protection Association publishes the *National Electrical Code® (NEC®)*[1], which is identified as ANSI/NFPA 70-1996 (the current edition).

The National Fire Protection Association is over 100 years old, having been organized in 1896. It has as its mission ''the safeguarding of man and his environment from destructive fire, using scientific and engineering techniques and education.'' The Association was incorporated in 1930 under laws of the Commonwealth of Massachusetts and has its headquarters in Quincy, Massachusetts.

NFPA is an independent, voluntary membership, nonprofit (tax-exempt) organization. A 30-member board of directors has general charge of the affairs of the Association, which has a staff of some 156 professional men and women, plus more than 144 support personnel.

NFPA is financed principally by sales of its publications and audiovisual materials, membership dues, income from seminars, research grants, and contributions. The annual budget for 1997 was about $56 million.

Membership in NFPA totals more than 69,000 individuals and 100 national trade and professional organizations. While the vast majority of NFPA's members are residents of the United States and Canada, the Association has members from over 70 other nations around the globe. Members are drawn from fire departments (24 percent), health care facilities (11 percent), business and industry (20 percent), the insurance field (6 percent), federal, state, and local government (7 percent), architecture and engineering (8 percent), fire equipment manufacturers and distributors (6 percent), trade and professional associations (2 percent), and other fields (16 percent).

1-3 NFPA Codes and Standards Development

NFPA codes and standards are prepared under the NFPA Regulations Governing Committee Projects. These regulations provide for a highly democratic method of preparation under a consensus process that permits wide public review and input into the system. Technical committees developing the codes and standards are required to be balanced so that not more than one-third of the membership of any technical committee represents a single interest.

Consensus is determined by requiring that at least two-thirds of the eligible voting members on a technical committee approve a change in an existing document or a new document before the change or new document is issued. Technical committee members are selected for their expertise in the particular subject. Many technical committee members represent national and even international organizations. Each member is classified as to one of the following interest categories:

Manufacturer
User
Installer/Maintainer
Labor
Applied Research and Testing
Enforcing Authority
Insurance
Consumer
Special Expert
Utility (*NEC* only)

The standards-making system provides for public proposals as well as proposals prepared by the technical committee, public review of the initial actions of the technical committee on the proposals and public comment on these actions, technical committee action on the comments, review by the NFPA membership, and action by the NFPA Standards Council, a subcommittee of the NFPA Board of Directors that has responsibility for administration of the NFPA standards-making system.

Provision is made for appeals to the Standards Council and in some cases for petitions to the NFPA Board of Directors when there is disagreement with the contents of a document or proposed document. Provision is also made for tentative interim amendments to the documents and for formal interpretations of the documents.

Figure 1-1 is a flow chart showing the development process for each new edition of the *National Electrical Code*. Other NFPA codes and standards are developed in the same manner, except they may not have a technical correlating committee overseeing the activities of the committee that developed the code or standard. NFPA recommended practices and guides are developed in the same manner.

1-3.1 The *National Electrical Code (NEC)*

The *National Electrical Code (NEC)* is only one of more than 285 codes, standards, guides, and recommended practices published by NFPA,

- Proposals received by NFPA and sent to technical committee (called a code-making panel on *NEC* committee).
- Technical committee meets and takes actions on proposals. Develops technical committee proposals.
- Technical committee letter balloted on actions at meeting. Reasons for negative votes recirculated to permit changes in voting.
- Technical correlating committee (if any) reviews results of balloting. Directs necessary actions as result of balloting and need for compliance with regulations and for correlation.
- Proposals and all actions published for public review and comment in *Report on Proposals (ROP).* For *National Electrical Code,* draft of the proposed new edition included with *ROP* to aid in understanding technical committee actions.
- Comments on *ROP* material received by NFPA and sent to technical committee.
- Technical committee meets and takes action on comments.
- Technical committee letter balloted on actions at meeting. Reasons for negative votes recirculated to permit changes in voting.
- Technical Correlating Committee (if any) reviews results of balloting. Directs necessary actions as result of balloting and need for compliance with regulations and for correlation.
- Comments and all actions published in *Report on Comments (ROC).* For *National Electrical Code,* advanced printing of new edition published to aid in understanding technical committee actions.
- NFPA members meet to review technical committee report in *ROP* and *ROC,* to act on motions at meeting to amend report, and to accept or return report to technical committee.
- Technical committee balloted on amendments passed at NFPA meeting and on suitability of document with changes due to amendments, etc., voted by members. Reasons for negative votes recirculated to permit changes in voting.
- Complaints (if any) on technical committee actions and membership actions sent to NFPA.
- Technical Correlating Committee (if any) reviews results of balloting and reviews any complaints. Makes recommendations to NFPA Standards Council on new or revised code or standard.
- Standards Council meets and acts on any appeals and on technical committee recommendation for new or revised code or standard.
- Petitions (if any) to NFPA Board of Directors received by NFPA.
- NFPA Board of Directors acts on any petitions.
- New or revised code or standard published by NFPA.

Figure 1-1. *Process for development of each new edition of the National Electrical Code.*

although it is by far the most widely adopted. The *National Electrical Code* provides the installation requirements for electrical equipment throughout most of the United States. According to the NFPA data base and the 1996 ''Directory of State Building Codes and Regulations,''[2] 45 states, Puerto Rico, the Virgin Islands, as well as cities, towns, and counties in all 50 states have electrical laws based on the *National Electrical Code.* It is also adopted by many government agencies. It has also been used as the basis for electrical regulation in several other countries.

The *National Electrical Code* was first issued in 1897. NFPA has sponsored the *Code* since 1911 and has published it since 1956. Like all other codes, standards, and other documents published by NFPA, it is processed in accordance with the NFPA Regulations Governing Committee Projects.

The National Electrical Code Committee consists of over 430 persons divided into 20 different code-making panels (technical committees) and a technical correlating committee. The code-making panels are responsible for specific articles in the *National Electrical Code.*

The *National Electrical Code* is organized into an introduction, nine chapters, and three appendices. The requirements of Chapters 1, 2, 3, and 4 apply generally and cover such subjects as definitions, general requirements for electrical installations, wiring design and protection, wiring methods and materials, and equipment for general use. Chapters 5, 6, and 7 apply to special occupancies (including hazardous locations), special equipment, or other special conditions. These latter three chapters supplement or modify the general rules in Chapters 1, 2, 3, and 4. Chapters 1 through 4 apply as amended by these chapters. Chapter 8 covers communications systems and is independent of the other chapters, except where they are specifically referenced in Chapter 8 articles. Chapter 9 consists of tables and examples. The appendices cover information on material extracted from other NFPA documents and reprinted in the *NEC*, application information on the formula for calculating conductor ampacities, and conduit and tubing conductor fill tables.

Each chapter of the *Code* is in itself divided into articles, parts, sections, and subsections. The articles covering hazardous locations are organized and arranged in a logical order to make it easier to find particular subjects.

Articles 500 and 505 cover the general requirements for electrical equipment and wiring for all voltages in locations where fire and explosion hazards may exist due to flammable gases or vapors, flammable liquids, combustible dust, or ignitible fibers or flyings.

Article 500 includes the definitions of hazardous locations classified by the long-standing United States ''Division'' classification system.

Article 505 includes the definitions of locations where explosion hazards may exist due to flammable gases and vapors or flammable liquids, classified by the long-standing international "Zone" classification system. This system is recognized in the United States as an alternate to the "Division" system of classification, with some restrictions.

Article 501 covers the requirements for Class I hazardous locations in general. Article 502 covers the requirements for Class II hazardous locations in general. Article 503 covers the requirements for Class III hazardous locations in general. Article 504 covers the installation requirements for intrinsically safe electrical systems. Articles 510 through 517 include requirements for specific hazardous locations, such as gasoline dispensing stations, commercial garages, aircraft hangars, and other special occupancies that are extremely common and for which the various NFPA technical committees have established specific guidelines for classification.

With a few exceptions, particularly in Article 503, Articles 501 through 503 are organized and numbered in the same way so as to facilitate location of the requirements. For example, the requirements for transformers and capacitors in Class I hazardous locations are covered in Section 501-2, for Class II locations in Section 502-2, and for Class III locations in Section 503-2.

The sections are subdivided into subsections or paragraphs (a), (b), and so forth. Subsection (a) of each section usually identifies the requirements for equipment and installations in Division 1 locations, and subsection (b) identifies the requirements for Division 2 locations. For example, Sections 501-2(a) and 502-2(a) give the requirements for transformers and capacitors in Division 1 locations, and Sections 501-2(b) and 502-2(b) give the requirements for Division 2 locations.

1-4 Other Organizations Publishing Standards on Hazardous Locations

There are a number of different organizations publishing standards on hazardous locations.

1-4.1 Underwriters Laboratories Inc.

Underwriters Laboratories Inc. (UL) publishes many different standards and proposed standards providing detailed construction and performance requirements for equipment for use in hazardous locations. Many UL standards are also recognized as American National Standards. These documents are used by UL to investigate the acceptability of equipment for use in hazardous locations and as a basis for permitting the UL listing mark or label on the equipment.

Copies of UL standards in print form, selected UL standards in electronic form or microfilm, and a current copy of the UL Catalog of Standards for Safety can be obtained by writing to Underwriters Laboratories Inc., Publications Stock, 333 Pfingsten Road, Northbrook, IL 60062-2096 (USA), telephone number (847) 272-8800 (ext. 42612 or 42622). The fax number is (847) 272-0293.

Single copies of UL standards for safety are also available from the UL offices in Melville, NY, telephone number (516) 271-6200 (ext. 22466 or 22301); Santa Clara, CA, telephone number (408) 985-2400 (ext. 2828), fax number (408) 296-3256; and Research Triangle Park, NC, telephone number (919) 549-1400 (ext. 1834), fax number (919) 549-1842.

The UL Catalog of Standards also identifies other organizations in the United States, and organizations in Australia, Canada, Germany, Italy, Japan, Korea, Mexico, the Netherlands, and the United Kingdom that can supply UL standards in print form, and for some suppliers, on microfilm, diskettes, or CD-ROM.

1-4.2 Factory Mutual Research Corporation

Factory Mutual Research Corporation (FM) publishes some standards and guides, although they primarily use the requirements of others in their investigation of equipment for use in hazardous locations. For more information, contact Factory Mutual Research Corporation, 1151 Boston-Providence Turnpike, Norwood, MA 02062 (USA), telephone number (781) 762-4300.

1-4.3 American Gas Association

The American Gas Association (AGA) publishes standards for gas-operated appliances. A copy of the AGA resource catalog can be obtained from the American Gas Association, 1515 Wilson Boulevard, Arlington, VA 22209 (USA), telephone number (703) 841-8588.

1-4.4 National Electrical Manufacturers Association

The National Electrical Manufacturers Association (NEMA) publishes standards on the construction and performance of electrical equipment. Where the construction and performance relate to safety, they usually reference the standards of others, such as Underwriters Laboratories Inc.

Information on NEMA standards can be obtained from the Publications Department, National Electrical Manufacturers Association, 1300 North 17th Street, Suite 1847, Rosslyn, VA 22202 (USA), telephone number (703) 841-3200.

1-4.5 American Petroleum Institute

The American Petroleum Institute (API) publishes standards relating to their industry and a catalog of publications and materials. Requests for the catalog and for publications should be directed to American Petroleum Institute, Publications and Distribution Section, 1220 L Street, NW, Washington, DC 20005 (USA), telephone number (202) 682-8375.

1-4.6 Occupational Safety and Health Administration

The United States Occupational Safety and Health Administration (OSHA) is a federal government regulatory agency with jurisdiction over certain occupancies within the United States. The requirements applied by OSHA are basically the requirements of the *National Electrical Code,* although the actual requirements are detailed in the applicable parts of the United States *Code of Federal Regulations*. See, for example, 29 CFR, Part 1910, Subpart S, Electrical, Section 1910.307 and Section 1910.399, Definition S.

1-4.7 Mine Safety and Health Administration

The United States Mine Safety and Health Administration (MSHA), another federal government regulatory agency, has regulations governing underground mining, including requirements for the construction and performance of equipment. For the regulations see the appropriate sections of the United States *Code of Federal Regulations*, 30 CFR, Parts 11 through 100.

1-4.8 United States Coast Guard

The United States Coast Guard (USCG) is another federal government agency with its own regulations governing hazardous locations published in the United States *Code of Federal Regulations*. Contact the Standards Development Branch of the Merchant Vessel Inspection Division, Office of Merchant Marine Safety at the United States Coast Guard, Commandant (G-MV1-2), USCG Headquarters, 2100 Second Street, SW, Washington, DC 20593 (USA).

1-4.9 Instrument Society of America

The Instrument Society of America (ISA) is another organization that publishes standards relating to equipment and installations in hazardous locations. Requests for the ISA Publications and Training Aids Catalog should be made to the Instrument Society of America, 67 Alexander

Drive, Research Triangle Park, NC 27709 (USA), telephone number (919) 549-8411.

1-4.10 Institute of Electrical and Electronic Engineers

The Institute of Electrical and Electronic Engineers (IEEE) is also an organization that publishes standards, some of which relate to equipment in hazardous locations. Requests for the IEEE Standards Catalog should be directed to the Institute of Electrical and Electronic Engineers Service Center, 445 Hoes Lane, Piscataway, NJ 08854 (USA), telephone number (908) 981-0060.

1-4.11 Foreign

There are a number of standards for the installation of equipment, classification of areas, and construction and performance of equipment that are published by organizations outside of the United States. For more details, see Section 6-1.

Bibliography

[1]NFPA 70, *National Electrical Code®*, National Fire Protection Association, Quincy, MA, 1996.

[2]"Directory of State Building Codes and Regulations," National Conference of States on Building Codes and Standards, Inc., Herndon, VA.

CHAPTER 2

General

2-1 What Hazardous Locations Are Not

In order to understand hazardous locations, it is first necessary to understand what they are not. See Section 2-3 for what hazardous locations are.

The simple presence or likelihood of the presence of a flammable or combustible material does not automatically classify an area as a hazardous location. For example, according to the definition of a Class I hazardous location (see Section 2-4.1), a single-family dwelling with a natural gas appliance installed could be classified as a hazardous location: the gas piping could leak, shutoff valves could fail, and so forth. There have, in fact, been a number of these types of failures, and even resulting explosions, in single-family dwellings. Some of the problems were the result of underground piping leaks outside the dwelling, with the gas entering the dwelling through the gas pipe or other underground opening in the foundation wall. In addition, even though a flammable natural gas-air mixture can and occasionally does occur within a single-family dwelling as a result of a gas leak, the ignition of this flammable mixture is usually from something other than an electrical source (such as a gas pilot light, smoking, and so on). Single-family dwellings with natural gas–fueled appliances are therefore not classified as hazardous locations, because the number of fires or explosions due to electrical equipment ignition of natural gas is small in comparison to the number of single-family dwellings, and requiring electrical equipment suitable for hazardous locations in single-family dwellings would do little to reduce what problem there may be.

2-1.1 Flash Point

Where a flammable or combustible liquid is present, it is necessary that the material be at or above its flash point before an explosion can occur.

According to NFPA 325, *Guide to Fire Hazard Properties of Flammable Liquids, Gases, and Volatile Solids*[1], the flash point of a liquid is the minimum temperature at which the liquid gives off vapor in sufficient concentration to form an ignitible mixture with air near the surface of the liquid within the vessel, as specified by the appropriate test procedure and apparatus.

It follows, therefore, that a flammable or combustible liquid must be at or above its flash point for an explosion to occur, and it is the prevention of explosions that is addressed by the *National Electrical Code*®[2] [except for Class III locations (see Section 2-6)].

Flammable and combustible liquids are classified based on the likelihood of explosive vapors being present. A flammable liquid is one having a flash point below 100°F (37.8°C) and having a vapor pressure not exceeding 40 psia (2068 mm Hg) at 100°F (37.8°C). A combustible liquid is one having a flash point at or above 100°F (37.8°C). For further details on the classification of liquids, see NFPA 30, *Flammable and Combustible Liquids Code*[3].

Ordinary ambient air conditions may result in the material being above its flash point. For example, aviation-grade gasoline has a flash point of −50°F (−46°C) and therefore emits explosive vapors at all but the most extreme low-temperature conditions.

Hydrogen is a gas at all but cryogenic temperatures and is therefore considered always hazardous. On the other hand, No. 1-D diesel fuel oil has a minimum flash point of 100°F (37.8°C) and therefore does not emit flammable vapors unless it is heated above this flash point. Most facilities handling No. 1-D diesel fuel oil would not be classified as hazardous locations. Exceptions might be in extremely hot locations or in areas where the diesel fuel oil is likely to come in contact with hot surfaces that could heat it above its flash point. However, even under these conditions, consideration must be given to the rapid cooling that occurs once the vapors move away from the hot surface. See Section 3-1.1.2 for more details.

See Appendix A, Table A-1, for a list of the flash points of some of the more common flammable and combustible liquids. More complete lists can be found in NFPA 325 and in other publications listing combustion properties. When the flash point cannot be determined through such references, tests can be conducted in accordance with appropriate ASTM standards, such as ASTM D 56[4], ASTM D 93[5], ASTM D 3278[6], ASTM D 3828[7], and ASTM D 1310[8]. The manufacturer of the material often lists this property in literature on the material.

2-1.2 Flammable (Explosive) Limits

All flammable gases and all flammable and combustible liquids heated above their flash points can be ignited. For most materials, there is both a minimum and a maximum concentration in air (oxygen) or other oxidizer below and above which propagation of flame does not occur when the gas/air mixture is in contact with a source of ignition. The mixtures are either too lean or too rich to be ignited. These concentrations are known as the lower and upper flammable or explosive limits. They are usually expressed in terms of percentage by volume of gas or vapor in air under normal ambient conditions.

The flammable limits change where air pressures or oxygen partial pressures are other than one normally encounters, or where other extreme conditions are present. The general effect of an increase in temperature or pressure is to lower the lower limit and to raise the upper limit. A decrease in temperature or pressure has the opposite effect. Changes in oxygen concentration, even under normal ambient temperatures and pressures, will have a marked effect on the flammable limits. Increasing the concentration will expand the range of concentrations that can be ignited. For additional information on the effect of oxygen enrichment, see NFPA 53, *Guide on Fire Hazards in Oxygen-Enriched Atmospheres*[9].

Some materials have a very wide flammable range (the spread between the lower and upper flammable or explosive limits), and some have a very narrow range. Acetylene, for example, has a lower flammable limit of 2.5 percent by volume in air, but the upper flammable limit is 100 percent. In other words, no air or oxygen is needed for acetylene to form an explosive mixture once its concentration exceeds 2.5 percent. Gasoline, on the other hand, has a relatively narrow range, approximately 1.4 percent to 7.6 percent, depending on its formulation and additives.

Processes can be and are designed to take advantage of the lack of an explosion hazard when the concentration is above the upper flammable limit. Extreme care is needed, however, to take advantage of this method of explosion protection: sometimes the mixture is within the flammable range (for example, during startup and shutdown). The most common approach is to maintain the concentration below the lower flammable limit. This can often be accomplished by adequate ventilation of the area. (See Section 3-1.1.2.)

See Appendix A, Table A-2, for a list of flammable limits of a few common flammable gases and flammable and combustible liquids. More complete lists can be found in NFPA 325 and in other publications listing combustion properties. Where such data is not available in the literature, ASTM E 681[10] is an existing standard method for the determination of flammable limits.

Combustible dusts also have flammable limits, usually called explosion concentrations, although they are not as well defined as those for liquids and gases, particularly in the upper limit. See Appendix A, Table A-3, for a list of minimum explosion concentrations for various dusts.

Even when present in the minimum explosion concentration, dust clouds are so thick it is impossible to discern objects more than three feet to five feet away. If you can see your hand in front of your face, the concentration probably is below the lower explosive limit.

Most of the studies in the United States on the properties of combustible dusts were done by the U.S. Department of Interior, Bureau of Mines. Much of this material, including information on the minimum explosion concentration of a large variety of combustible dusts, appears in NMAB 353-4[11].

2-1.3 Pyrophorics and High Explosives

The presence of pyrophoric materials (materials that ignite spontaneously in contact with air), such as some phosphorus compounds, very finely divided zirconium powder, and some aluminum alkyl solutions, or the presence of high explosives, such as blasting agents and dynamites, does not mean that an area is classified as a hazardous location in accordance with the definition in the *National Electrical Code*. Where pyrophoric materials are present, the danger of ignition from electrical energy is much less than it is from other sources. However, the established requirements for areas handling such materials often include requirements for electrical equipment suitable for use in hazardous locations. Regulations governing the handling of munitions often require electrical equipment suitable for use in hazardous locations. The assumption, apparently, is that if the equipment is suitable for Class II, Division 1 locations, the dust from high explosives in handling and manufacturing areas will not enter dust-ignitionproof enclosures, and therefore safety is increased.

In addition, the surface temperature of equipment suitable for use in hazardous locations is limited. Furthermore, explosionproof and flameproof enclosures are strong enough to withstand explosions within them without propagating the explosions to the outside, under the test conditions specified for such equipment.

This is sound engineering up to a point. The problem is that equipment listed or approved for use in hazardous locations as defined in the *National Electrical Code* has been investigated for use only with certain materials, not including munitions. The use of explosionproof, flameproof, and dust-ignitionproof enclosures certainly increases the level of safety beyond that which would be available should the enclosures be "ordinary location" enclosures.

2-2 Hazardous vs. Classified Locations

Hazardous locations are sometimes referred to as classified locations. Article 500 of the *National Electrical Code* uses the term "Hazardous (Classified) Locations" as its title. Some codes use one term, some the other, and some both. It has been argued that all areas described in this book as ''hazardous locations'' should be described as ''classified locations'' or ''classified areas,'' because they have been classified. Some interests prefer not to use the term ''hazardous,'' and other interests prefer not to use the term ''classified.'' As used in NFPA codes and standards and in the *National Electrical Code*, the terms are interchangeable.

Hazardous (classified) locations, if properly treated, are not necessarily any more dangerous to work in than other areas or locations. Many more people are killed in traffic accidents and general industrial accidents than are killed as the result of explosions in hazardous locations. Furthermore, hazardous locations are not hazardous to health if proper precautions are taken, although most flammable and combustible materials (some exceptions are hydrogen and methane) are hazardous to health if released into the atmosphere in sufficient concentrations, often far below the lower flammable limit.

The term ''hazardous'' is maintained in the *National Electrical Code* because it provides, by its very name, a degree of warning. In some countries the term ''explosion endangered'' has been used. Internationally, the term ''explosive gas atmospheres'' is used.

The terms ''location'' and ''area'' are also used interchangeably, and in some standards the terms ''space'' or ''spaces'' is used. Whatever the term used, it is intended to encompass a three-dimensional space, not a flat (that is, two-dimensional) area.

2-3 What Hazardous Locations Are

Hazardous locations are those locations, areas, or spaces where fire or explosion hazards may exist due to flammable gases or vapors, flammable liquids, combustible dust, or ignitable fibers or flyings. As used here, the hazard is an explosion hazard when flammable gases or vapors, flammable liquids, or combustible dusts are present (Class I and Class II locations). The fire hazard relates to the presence of ignitible fibers or flyings (Class III locations).

The term ''vapors'' as used in the *National Electrical Code* and in this book means flammable gases emitted from the surface of flammable or combustible liquids. See the information on flash point in Section 2-1.1. This is a nonscientific but common way of expressing the term. One does not normally refer to ''gasoline gases''; one calls them ''gasoline vapors.'' However, the effect is the same.

Materials do not have to be in a gaseous state for an explosion to occur. Combustible dusts are an obvious example. Atomized liquids, even below their flash point, burn extremely rapidly. The finer the droplet size, the higher the flame speed. Flame speeds (flame velocities or burning rates) of materials vary with the material and material concentration. The higher the flame speed, the more ''violent'' any resultant explosion; that is, the higher the explosion pressure and rate of pressure rise. Ignition of very finely atomized liquids below their flash point can result in flame speeds that approach those associated with the explosion of the same material in the gaseous state.

2-4 Class I Locations

Class I locations involve flammable and combustible liquids and flammable gases.

2-4.1 *National Electrical Code* Definition

The *National Electrical Code* defines a Class I location as a location in which flammable gases or vapors are or may be present in the air in quantities sufficient to produce explosive or ignitible mixtures. It includes those locations defined as Class I, Division 1; Class I, Division 2; Class I, Zone 0; Class I, Zone 1; and Class I, Zone 2. (See Section 2-4.3.)

2-4.2 *NEC* Article 500 Groups

See Section 2-4.2.6 for *NEC* groups.

2-4.2.1 Why They Exist and What They Are

Groups exist to permit classification of locations depending on the properties of the flammable vapors, liquids, or gases, and to permit testing and approval of equipment for such locations.

Until publication of the 1937 edition of the *National Electrical Code*, Class I hazardous locations were not divided into groups. All flammable gases and vapors were classified as a single degree of hazard. It was recognized, however, that the degrees of hazard varied, and that equipment suitable only for use where gasoline was handled was not necessarily suitable for use where hydrogen or acetylene was handled.

It was also recognized that it was very difficult to manufacture equipment and enclosures for use in hydrogen atmospheres, and that the equipment, even if built, was expensive. It was not logical from an engineering standpoint to require explosionproof equipment in gasoline filling stations that was also suitable for use in hydrogen atmospheres. Not only would this unnecessarily increase the cost of the electrical installation in one of the most common types of hazardous locations, but it would make some types of equipment unavailable. Even today, there are no listed explosionproof motors or generators suitable for use in Group B or A atmospheres.

The solution was to divide Class I hazardous locations into groups, with each group containing materials of similar explosion characteristics. This permitted explosionproof equipment to be built that was no more expensive than necessary for the particular application. The dividing lines for the groups were chosen on the basis of the then most common commercial materials: acetylene, hydrogen, ethyl ether, and gasoline.

In the 1937 edition of the *NEC* the definitions of the Class I groups were as follows:

Group A: Atmospheres containing acetylene

Group B: Atmospheres containing hydrogen or gases or vapors of equivalent hazard, such as manufactured gas

Group C: Atmospheres containing ethyl ether vapor

Group D: Atmospheres containing gasoline, petroleum, naphtha, alcohols, acetone, lacquer solvent vapors, and natural gas

Over the years this list of materials expanded somewhat as more test information became available and as the need for classification of additional materials became evident.

By the 1968 edition of the *NEC*, the groups had been expanded as follows:

Group A: Atmospheres containing acetylene

Group B: Atmospheres containing hydrogen, or gases or vapors of equivalent hazard, such as manufactured gas

Group C: Atmospheres containing ethyl ether vapors, ethylene, or cyclopropane

Group D: Atmospheres containing gasoline, hexane, naphtha, benzine, butane, propane, alcohol, acetone, benzol, lacquer solvent vapors, or natural gas

In 1969 Underwriters Laboratories Inc. published Bulletin of Research No. 58[12]. The research was undertaken at the suggestion of the National Fire Protection Association's Sectional Committee on Electrical Equipment in Chemical Atmospheres (now the NFPA Technical Committee on Electrical Equipment in Chemical Atmospheres) with the cooperation and financial sponsorship of the American Insurance Association, the American Petroleum Institute, the Manufacturing Chemists Association (now the Chemical Manufacturers Association), and the National Electrical Manufacturers Association. The work included development of special test equipment described in more detail in this chapter. (See Figure 2-4.)

As a result of the UL research a tentative interim amendment was issued for the 1968 edition of the *National Electrical Code*, and the 1971 edition of the *NEC* was expanded to include a number of additional chemical materials in Groups B, C, and D. Additional research resulted in further expansion of the list of chemicals into groups in the 1981 edition of the *NEC*. In that edition, 45 different materials were identified in Group D alone.

Because of this ever-expanding list of chemicals, the National Electrical Code Committee decided to specify for the 1984 edition of the *NEC* only the major groups of chemicals, as they had done in the past. In 1983 the NFPA Committee on Electrical Equipment in Chemical Atmospheres, which had responsibility for the classification of the chemicals into groups in accordance with the NFPA standards-making system, published the first edition of NFPA 497M, *Manual for Classification of Gases, Vapors, and Dusts for Electrical Equipment in Hazardous (Classified) Locations*[13]. This manual, which was referenced in the 1984 *NEC*, included the expanded list of chemicals. This change permitted the NFPA committee responsible for the subject to have direct control of the list and to revise it annually, if necessary, instead of every three years as new editions of the *NEC* are published. The material in NFPA 497M has recently been included in NFPA 497, *Recommended Practice for the Classification of Flammable Liquids, Gases, or Vapors and of Hazardous (Classified) Locations for Electrical Installations in Chemical Process Areas*[14], and NFPA 499, *Recommended Practice for the Classification of Combustible Dusts and of Hazardous (Classified) Locations for Electrical Installations in Chemical Process Areas*[15]. NFPA 497M has been discontinued.

2-4.2.2 How They Are Determined

Prior to publication of the 1971 edition of the *National Electrical Code*, the determination of the proper group classification (as defined in Article 500) in Class I locations was based on the following three criteria:

1. Maximum experimental safe gap (MESG)
2. Explosion pressure
3. Ignition temperature

2-4.2.3 Maximum Experimental Safe Gap

It has been known since before electric power became commercially available in the late nineteenth century that ignition of a flammable atmosphere could be prevented by separating the unburned flammable atmosphere from the ignition source by a screen or gauze material with very fine openings. This principle was used in underground mining operations where methane gas was a constant hazard. Miners carried lamps, in which the illuminating source was a flame, into gassy mines. Ignition of the surrounding flammable methane-air mixture did not occur, because the flame was surrounded by fine metal gauze. This is the principle of the Davy lamp, invented in 1815 by Sir Humphry Davy.

(See Figure 2-1.) This lamp also served to detect the presence of a flammable atmosphere: when methane was present, the gas within the gauze next to the flame was ignited, creating an elongated flame inside the gauze. The Davy lamp is still in use in some mines.

Subsequent research using holes in flat metal plates showed that for each flammable material, there was a maximum opening dimension through which a burning mixture on one side of the opening would not propagate a flame to a flammable mixture on the other side of the opening. This principle was used when electricity was introduced into underground mines. Electrical equipment, such as switches and motors with commutators, was enclosed in such a way that switch operating mechanisms and openings in the enclosure for the rotating motor shafts

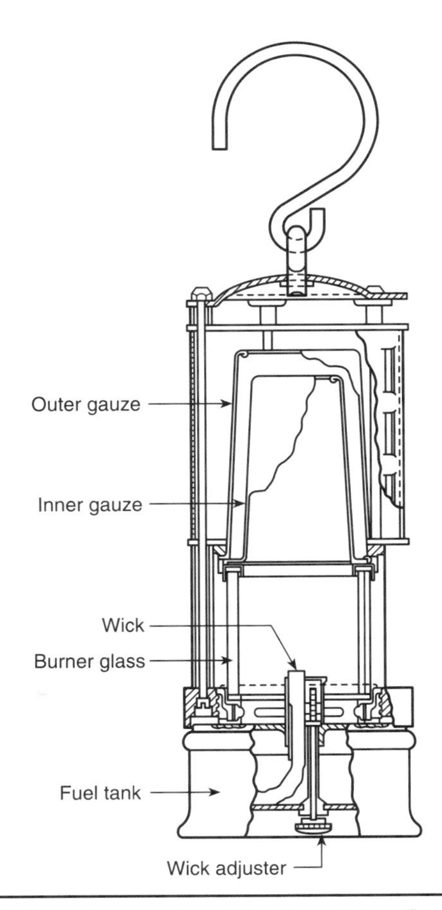

Figure 2-1. *A modern version of the Davy lamp. Two fine-gauze screens separate the flame of the lamp from the surrounding methane-air mixture. (Courtesy of Koehler Mfg. Co.)*

were very small. It was found that flat joints having a very small opening or gap between mating surfaces in comparison to their width presented a path such that, even though the explosive mixture was inside the enclosure and was ignited inside the enclosure, propagation of the explosion to the flammable mixture outside the enclosure was prevented. This path, known as the flame path, has been the subject of considerable research in an effort to ascertain how it can be determined based on the chemical and combustion properties of the flammable material involved. Such a method would be extremely valuable, as it would permit at least the major part of the determination of the proper Class I hazardous location group without costly and time-consuming testing.

In the meantime, a number of different test methods have been developed to determine this safe gap between mating services by explosion test experiment. The gap, thus determined, is known as the maximum experimental safe gap, or MESG.

In building explosionproof equipment (the definition of explosionproof apparatus is given in Section 4-2.1.1), it is assumed that a flammable mixture will enter the enclosure. It can do this readily through shaft and other openings, such as cover-to-body joints, conduit and cable fitting entries, and as a result of "breathing" of the apparatus from heating and cooling cycles (typical of motors and lighting fixtures). The intent, is twofold: to construct the enclosure so that it is strong enough to withstand the anticipated internal explosion, and to prevent the resulting flames or hot gases from igniting the surrounding external atmosphere.

The process of preventing the flame and expanding hot gases within an explosionproof enclosure from igniting the external surrounding flammable atmosphere is often explained as a process of the gases cooling as they pass through the enclosure joints. However, the process is far more complicated than this. When the hot gases are ejected from the enclosure through the small gap between the mating surfaces, they come out as a high-speed (sometimes supersonic) hot jet. (See Figure 2-2.)

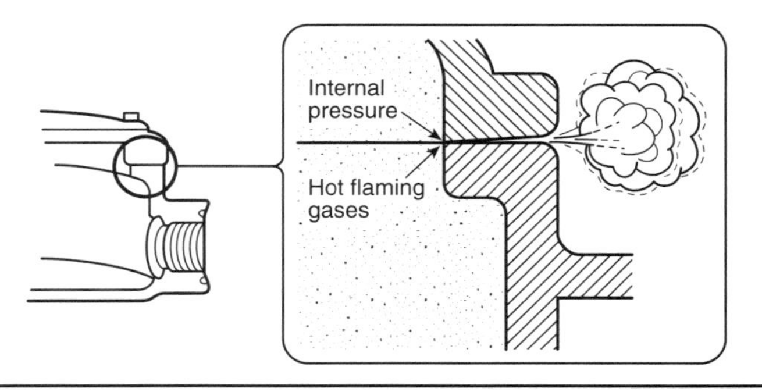

Figure 2-2. *Hot gases being ejected from an explosionproof enclosure joint.*

This jet is cooled rapidly by the surrounding flammable atmosphere and the turbulence created once it leaves the enclosure. It is this mixing and cooling phenomenon under turbulent conditions that makes it so difficult to predict the MESG for any particular flammable material. The size and shape of the gap, the pressure that forces ejection of the material, the time period over which the hot gases are ejected, the temperature of the hot gases, the degree of turbulence, and the thermal conductivities of the mixtures are all critical in the determination.

The pressure forcing the jet out of the enclosure through the gap is related to the composition of the burning mixture inside the enclosure, the shape of the enclosure, and the volume of material, as well as the point and energy of ignition inside the enclosure. The time period is related to the volume of material and its burning rate.

The shape of the jet of hot gases is related to the configuration of the joint under the explosion conditions. Testing equipment designed to determine the MESG is constructed so that the shape of the gap and the distance between mating surfaces will not change as a result of the explosion. Such a condition is unlikely to exist in practice, however. Even heavy cast-iron enclosures with flat joints that are bolted together change shape during the dynamic process of an explosion inside the enclosure. (See Figure 2-3.)

The greater the pressure for any given enclosure, the larger the gap between bolts. Even though the gap is smaller than the MESG when

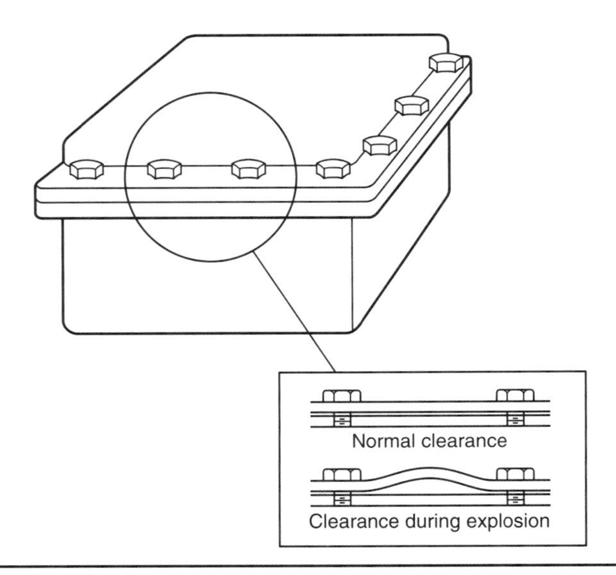

Figure 2-3. *Joint clearances in explosionproof enclosures change under explosion conditions.*

measured with a feeler gauge, it may well be larger under actual explosion conditions and thus permit propagation of the explosion within the enclosure to the surrounding flammable atmosphere.

Method Used by Underwriters Laboratories and the *National Electrical Code* for Determining MESG: The method used for classification of Class I materials into the groups defined in Article 500 of the *National Electrical Code* and used by the NFPA Committee on Electrical Equipment in Chemical Atmospheres is the one developed by Underwriters Laboratories Inc. and described in detail in the UL Bulletin of Research No. 58. The methods used for classification into the groups defined in Article 505 of the *NEC* are somewhat different. The Article 500 method is designed to compare the performance of materials, one to the other, by determining the MESG between adjacent cylindrical chambers, both filled with the same flammable mixture, under the following conditions:

1. Quiescent conditions with ignition adjacent to gap
2. Quiescent conditions with ignition away from gap
3. Quiescent conditions with ignition at end of a 1½ in. trade size by 10-ft (3.05-m) length of pipe, representing rigid conduit
4. Turbulent conditions

The equipment is known as the "Westerberg Explosion Test Vessel," named for W. C. Westerberg, the UL engineer who designed it. (See Figure 2-4.)

The explosion is initiated in a primary explosion chamber, which is about 1 ft^3 (28.32 L) in volume. An adjustable rigid gap about 4 in. (102 mm) long and adjustable from ⅜ in. (9.52 mm) to 1⅛ in. (28.6 mm) wide is set between this chamber and a somewhat smaller secondary chamber. Both chambers are equipped with explosion-pressure sensing transducers and a sensitive pressure gauge for determining internal pressures during setup. A horizontal length of 1½-in. trade size pipe representing rigid conduit, 10 ft (3.05 m) long, is attached to the back of the primary chamber with a spark gap at the far end. A valve system permits this pipe to be isolated from the primary explosion chamber. A blower system is installed in the primary chamber so that turbulent conditions can be created, similar to the conditions inside a motor and as an aid in mixing the gas- or vapor-air mixture. Spark gaps are installed opposite the gap opening close to the gap, and near the back of the chamber away from the gap.

The secondary chamber is equipped with an observation port so that the results can be observed visually or by a closed-circuit television camera connected to a video tape recorder.

Figure 2-4. *UL's Westerberg explosion test vessel.*

The gap between the two chambers is first adjusted to a known clearance and checked with feeler gauges. The next step is to evacuate both chambers and the pipe, if connected, to an extremely low pressure, then to inject a known amount of the flammable gas or liquid involved. A heating and carburetor system is provided to permit vaporization of liquids with a flash point above room ambient. The ambient air is then allowed to enter both chambers and the pipe, if used, until the internal pressure is the same as the external atmospheric pressure. The percent by volume of flammable material in air is calculated by the partial pressures method. The flammable mixture is thoroughly mixed between the two chambers and the pipe, if connected, by a valving arrangement and the blower motor.

The mixture in the primary chamber or pipe is then ignited by an electric spark, and a determination is made of the instantaneous pressures in both chambers as well as whether or not ignition has occurred in the secondary chamber. This process is repeated for each condition of ignition and through a variety of concentrations of the flammable mixture to determine the following:

1. The maximum gap under each condition of ignition, turbulence, and so on, through which the mixture will not propagate between chambers
2. The maximum explosion pressure in both primary and secondary chambers under each test condition

These data are then compared to the base data for the four dividing line materials in Table 2-1.

If the unknown material being tested has an MESG under any test condition greater than 0.029 in. (0.737 mm), it is tentatively classified Group D, because 0.029 in. (0.737 mm) is the MESG of gasoline (Naphtha R) under the same test conditions.

If the unknown material has an MESG of between 0.012 and 0.029 in. (0.305 and 0.737 mm), it is tentatively classified Group C, because 0.012 in. (0.305 mm) is the MESG of ethyl ether (Group C).

If the unknown material has an MESG of less than 0.012 in. (0.305 mm), but not less than 0.003 in. (0.076 mm), it is tentatively classified either Group A or Group B, depending upon the explosion pressure.

If the unknown material has an MESG of less than 0.003 in. (0.076 mm), it is identified as a material incapable of classification by Article 500 of the *NEC*. Carbon disulfide is one such material, with an MESG of 0.002 in. (0.051 mm).

Explosion Pressure: After the initial tentative classification, the classification is reviewed again based on recorded explosion pressures. For example, explosion pressures for materials tentatively classified Group D are compared to the explosion pressures for the basic comparison material, gasoline. If the explosion pressures are approximately the same or less than those observed with gasoline, classification as a Group D material is confirmed. If they are higher, consideration is given to classification into a more dangerous Group (C, B, or A) because of the higher explosion pressure.

In general, it has been found that materials have higher explosion pressures and lower MESGs under turbulent testing conditions than

Table 2-1 Dividing Line Between Groups

Group	Dividing Line Material	Group
A	Acetylene	B
B	Hydrogen	C
C	Ethyl ether	D
D	Gasoline*	

**Actually a rubber solvent representing unleaded gasoline, known as "Naptha R."*

under quiescent conditions. It has also been found that the MESG is usually lowest with ignition adjacent to the gap and that explosion pressures are sometimes considerably higher with ignition at the end of a length of pipe.

This phenomenon of higher explosion pressures with a length of pipe (representing rigid conduit) is commonly referred to as "pressure piling." It can be attributed to the prepressurization of the unburned mixture ahead of the moving flame front, although this is an oversimplification of the actual process.

Some materials exhibit this characteristic of pressure piling to a much greater degree than others — some to the extent that they are dual-rated materials in the *NEC.* A dual-rated material is one with an MESG that places it into one group, but when tested under pressure-piling conditions with a length of simulated rigid conduit, it exhibits an explosion pressure that puts it into a more dangerous group.

The material butadiene is one such material. Butadiene has a MESG of 0.031 in. (0.787 mm), well within the Group D range. However, under pressure-piling conditions with ignition at the end of a length of pipe, the explosion pressure is between that observed with ethyl ether (Group C) and that observed with hydrogen (Group B), thus putting it into the Group B classification. The material is therefore classified Group B, with a provision that if all conduit entries are sealed, equipment suitable for Group D locations can be used. Sealing of all conduits, even $^1/_2$-in. trade size, reduces the likelihood of pressure piling.

Other materials that are dual rated are ethylene oxide, propylene oxide, and acrolein. In these cases, however, the materials are classified as Group B materials, and Group C equipment can be used if all conduit entries are sealed.

Although hydrogen and acetylene have the same MESG, acetylene produces considerably higher explosion pressures and is therefore in its own category, Group A.

For the group classifications of common materials, see Appendix A, Table A-4.

National Research Council Work: The Committee on Evaluation of Industrial Hazards of the National Materials Advisory Board, Commission on Engineering and Technical Systems, National Research Council, National Academy of Sciences, and their predecessor committees worked for a number of years in an attempt to develop a method for classification of Class I materials based on methods other than explosion tests. These committees used the facilities of Underwriters Laboratories Inc. to test a variety of materials, including materials representing different chemical families and believed to be borderlines between different hazardous location groups. The details of the UL tests are in UL Bulletins of Research Nos. 58A[16] and 58B[17].

The National Research Council committees were able to develop a rationale for classification of materials that was based on the chemical "family" of the material. For complete details, see NMAB 353-6.[18] As with all other methods (other than testing) developed to date, it does not work for all materials.

Additional work has been done outside of the United States to determine the proper classification into groups based on minimum igniting current. It is one of the methods used for classification of materials into the groups defined in Article 505 of the *NEC*. This minimum igniting current method provides good guidelines as far as group classification is concerned, but it does not take into account changes in classification because of unusual explosion pressure conditions.

Differences Based on Test Method: The National Research Council committee and others have recognized that the test method used to determine the MESG may have a bearing on the results. This difference was investigated in some detail and reported on by Strehlow et al.[19]

Other countries and organizations have developed test methods for determining the group classification of flammable gases and vapors. These test methods usually involve spherical explosion test vessels of various sizes [1.22 in.^{3} (20 ml) and 488 in.^{3} (8 L) are the two most common sizes] with circumferential gaps. Since the countries that use these methods do not usually use conduit wiring systems, they are not concerned with the pressure-piling problems associated with such wiring systems and therefore do not use lengths of pipe in their testing apparatus. Also, the testing equipment does not provide for testing turbulent mixtures. On the other hand, some of this test equipment has provision for separate control of the concentration of the mixture inside and outside the test chamber, which is not present in the UL test apparatus. A concentration that is the most easily ignited is not always the same as the concentration that is the most easily propagated through a gap.

One major problem with the original UL test apparatus just described is that the secondary chamber is smaller than the primary chamber, so that the mixture in the secondary chamber is often pressurized before it is ignited by a flame or hot gases passing through the gap from the primary chamber. This problem is overcome by the construction of the test chambers used by other countries that have the secondary chamber, in effect, infinite in volume in comparison with the primary chamber. This difference is explored in some detail in the "Anomaly" paper, and in a Canadian paper by Brown et al[20]. UL has developed a modification of the Westerberg apparatus designed to overcome the problem.

For additional information on classification of gases and vapors into groups, see NMAB 447[21].

2-4.2.4 Ignition Temperature

The ignition temperature of a flammable gas or vapor is critical in determining the acceptability of equipment that operates at relatively high external surface temperatures, such as lighting fixtures, heaters, and motors. Explosionproof equipment is designed to contain an explosion and prevent propagation of the explosion to the surrounding flammable atmosphere. However, if the external surface of the explosionproof enclosure is at a temperature above the ignition temperature of the flammable gas or vapor, the external surface can act as an ignition source in itself, defeating the intent of the protective enclosure.

High external surface temperatures can be created by the normal operation of the electrical equipment, such as an incandescent lamp bulb in a lighting fixture; by the abnormal operation of the equipment, such as overheating of a motor when the rotor stalls; or because of a fault condition, such as an arcing fault to the enclosure.

By definition, the ignition temperature of a substance, whether solid, liquid, or gaseous, is the minimum temperature required to initiate or cause self-sustained combustion independently of the heating or heated element. It is sometimes (and more correctly) referred to as "autoignition temperature" or "apparent ignition temperature" (AIT). For more detailed information, see NFPA 325.

Ignition temperatures observed under one set of conditions may vary substantially by a change of conditions, including a change in the test method. For this reason, ignition temperatures should be looked upon only as approximations. Some of the variables known to affect ignition temperatures are percentage composition of the vapor- or gas-air mixture, the shape and size of the space where the ignition occurs, the rate and duration of heating, the kind and temperature of the ignition source, catalytic or other effects of materials that may be present, and oxygen concentration.

Much of the data for the determination of ignition temperature that have been reported in various publications have been obtained by one of two standard procedures: ASTM D 286[22] or ASTM D 2155[23]. Both standards have now been withdrawn by ASTM: ASTM D 286 many years ago and ASTM D 2155 as of November 1980. ASTM D 2155 has been replaced by ASTM E 659[24].

The various test methods differ in the following basic ways:

1. The size and shape of the test vessel
2. The method of heating the test vessel
3. The method of detecting ignition

In all test methods, the test vessel is a borosilicate glass flask. The test vessel is heated thoroughly in an insulated enclosure to a known temperature. A carefully measured amount of the material to be tested is injected into the test vessel. If no ignition occurs, the test vessel temperature is raised and the test is repeated. The test is also repeated using various concentrations until the lowest temperature of ignition for any concentration is found. (See Figure 2-5.)

Increasing the volume of the test vessel normally results in a lowering of the observed ignition temperature. Changing the shape of the test vessel (surface-to-volume ratio) also affects the temperature observed, as does the material of the test vessel. Borosilicate glass has been found to result in the lowest temperatures of any material that does not result in a catalytic action.

Since the tests are designed to heat the entire flammable mixture in its most easily ignited concentration, there is a degree of safety in the test. In most installations of electrical equipment, the flammable mixture will be heated as it contacts the hot surface, and turbulence at the surface will result so that the flammable mixture is never as hot as the heated surface.

On the other hand, if the flammable mixture is in a closed, heated chamber, such as a drying oven, which is much larger in volume than the test vessel, ignition can occur at lower than the recorded ignition temperature.

Prior to publication of the 1971 edition of the *National Electrical Code,* the ignition temperature of the flammable material was part of the group classification process. Equipment intended for Group D locations was limited to a maximum surface temperature of 280°C (536°F), as was equipment for Groups A and B. However, equipment for use in Group C was limited to an external surface temperature of 180°C (356°F), the ignition temperature of ethyl ether at that time. Subsequent tests indicated an ignition temperature for ethyl ether of 160°C (320°F). Thus, a new material being investigated for classification, even though it might be classified as a Group D material because of explosion pressure and MESG considerations, would have to be classified as Group C if the ignition temperature of the material was less than 280°C (536°F). If the material had an ignition temperature less than 180°C (356°F), it could not be classified. Carbon disulfide, with an ignition temperature of 90°C (194°F), is one such material.

This problem was recognized by the National Electrical Code Committee, and the 1971 edition of the *NEC* was revised to remove the ignition temperature as a criterion for group classification. A system of marking equipment to identify the external surface temperature was instituted in its place, and a requirement was established that equipment could not be used in locations where the ignition temperature of the flammable material was less than the marked external surface temperature of the equipment.

A system of identification numbers was also established, giving

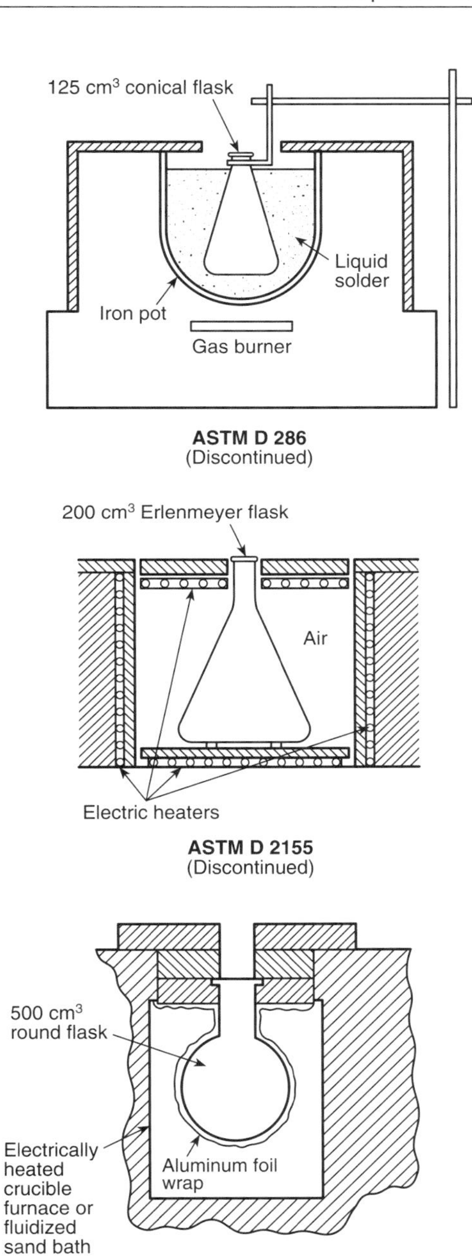

Figure 2-5. *Test apparatus for determining ignition temperature.*

specific temperatures or temperature ranges. This system was based on the European and International Electrotechnical Commission (IEC) system, with additional levels based on the traditional dividing lines in the United States, as shown in Table 2-2.

Thus, based on the international level of 300°C (572°F), the identification number is T2. There was no international level at a temperature of 280°C (536°F), one of the traditional dividing lines in the United States, so a new identification, T2A, was established. This permitted equipment used in the United States to be marked T2A. The equipment also complied with the T2 marking traditionally used in Europe, since the temperature did not exceed 300°C (572°F).

See Appendix A, Table A-5, for the ignition temperatures of some common materials.

2-4.2.5 Where to Go for Information on Unknown Materials

National Fire Protection Association Materials: There are several NFPA publications that include lists of chemicals. NFPA 49, *Hazardous Chemicals Data*[25], is intended primarily to provide information on fire and explosion hazards, life hazards, personal protection, fire fighting, shipping, and storage. For some very common chemicals, such as ammonia, it identifies the Class I hazardous location group.

NFPA 325 includes information on the flash point, ignition temperature, flammable limits, specific gravity, vapor density, boiling point, extinguishing methods, and hazard identification in accordance with

Table 2-2 External Surface Temperature Identification Numbers

Maximum Temperature °C	Maximum Temperature °F	Identification Number
450	842	T1
300	572	T2
280	536	T2A
260	500	T2B
230	446	T2C
215	446	T2D
200	392	T3
180	356	T3A
165	329	T3B
160	320	T3C
135	275	T4
120	248	T4A
100	212	T5
85	185	T6

NFPA 704, *Standard System for the Identification of the Hazards of Materials for Emergency Response*[26].

The 1997 edition of NFPA 497 includes information on the group identification and the ignition temperature. This information was previously found in NFPA 497M, which is no longer published.

National Research Council Data: Much of the classification data in NFPA 497 (and the old NFPA 497M) is based on work done by the Committee on Evaluation of Industrial Hazards, the National Materials Advisory Board, the Committee on Engineering and Technical Systems, the National Research Council, and their predecessor committees. This data is published in NMAB 353-5[27].

Other Sources: The material manufacturer should always be consulted for information on the proper group classification of an unknown material. It may also be possible to provide some idea of the proper group classification of materials using the concepts for group classification developed by others. See the previous paragraph on the work by the National Research Council. In some cases, it may be possible to have tests conducted by qualified testing laboratories with the proper equipment.

2-4.2.6 *NEC* Article 505 Groups

Article 505 of the *National Electrical Code*, which is new in the 1996 edition, recognizes the international system of area classification as an alternate to the Article 500 system, with restrictions. This classification system is described in detail in the International Electrotechnical Commission (IEC) Publication 79, Part 10[28]. This and other IEC publications are available at Bureau Central de la Commission Electrotechnique Internationale, 3 rue de Verembé, Genève, Suisse. IEC publications are also available from the American National Standards Institute (ANSI), 11 West 42nd Street, 13th floor, New York, NY 10036, as well as some U.S. and international standards libraries.

2-4.2.6.1 Why They Exist and What They Are

Although the background on the development of the IEC classification differs from the *NEC* Article 500 system, the groups exist for exactly the same reasons, as described in Section 2-4.2.1. Since Group I is reserved in the IEC system for classification and equipment in underground mines, outside the scope of the *NEC*, all groups are identified with the Roman numeral II followed by a letter, A, B, or C.

In Article 505 the materials in the groups are as shown in Table 2-3.

Table 2-3 Grouping of Materials per Article 505

Group	Materials in Group
IIC	Acetylene, hydrogen, or gases or vapors of equivalent hazard
IIB	Acetaldehyde, ethylene, or gases or vapors of equivalent hazard
IIA	Acetone, ammonia, ethyl alcohol, gasoline, methane, propane, or gases or vapors of equivalent hazard

Two facts are immediately evident when comparing the groupings of Article 505 in the table to the groupings defined in *NEC* Article 500:

1. The lettering of the groups (A, B, C) is reversed.
2. There are only three groups in the Article 505 system, instead of four as in the Article 500 system.

In Article 500, the most dangerous material, acetylene (it has the highest explosion pressure and the smallest MESG) stands alone in Group A, and the materials with the lowest explosion pressures and largest MESGs are in Group D. In Article 505, the most dangerous materials are in Group IIC and the least dangerous are in Group IIA. This is the result of the long-standing histories of the two classification systems. When adopting the IEC system described in Article 505, the *NEC* Committee judged it to be more confusing, and in fact dangerous, to try to change either identification system, particularly in view of the marked equipment available on the market.

In the Article 505 system, acetylene is grouped with hydrogen as it is in the IEC system, thereby eliminating one of the four Article 500 groups. Otherwise, all materials classified in Group IIA in Article 505 are in Group D in Article 500. All materials in Group IIB in Article 505 are in Group C in Article 500. And all materials in Group IIC in Article 505 are in Group B and A in Article 500. Table 2-4 compares the two groups.

2-4.3 Divisions and Zones

2-4.3.1 Why They Exist and What They Are

The Class I hazardous location division or zone identifies the likelihood of a flammable mixture being present. This varies all the way from

Table 2-4 Comparison of Grouping, Articles 505 and 500

Article 505 Group	Equivalent Article 500 Group
IIC	A and B
IIB	C
IIA	D

continuously to never. If it is never present, the area is not a hazardous location. The more likely a flammable mixture will be present, the more likely an explosion.

From an engineering standpoint, greater precautions are needed if a particular set of circumstances is likely to occur (such as the presence of a flammable mixture within the explosive range) than are necessary if those circumstances are unlikely to occur. This is the reason for dividing hazardous locations into two divisions per *NEC* Article 500 or three zones per Article 505.

2-4.3.2 Divisions

Divisions are defined in *NEC* Article 500, whereas zones are defined in Article 505. This has led to the terms "Division System" and "Zone System" when referring to the two systems of area classification.

2-4.3.2.1 Division 1

Class I, Division 1 hazardous locations are defined in the *National Electrical Code* as follows:

1. Those locations in which ignitible concentrations of flammable gases or vapors can exist under normal operating conditions
2. Those locations in which ignitible concentrations of such gases or vapors may exist frequently because of repair or maintenance operations or because of leakage
3. Those locations in which breakdown or faulty operation of equipment or processes might release ignitible concentrations of flammable gases or vapors and might also cause simultaneous failure of electric equipment

Note that in each case, "ignitible concentrations" are mentioned. This means concentrations between the lower and upper flammable or explosive limits.

The *National Electrical Code* definition is followed by a fine-print note (explanatory material) that describes a number of areas and occupancies normally classified as Class I, Division 1 locations. It uses the phrase "volatile flammable liquids." These are specifically defined in Article 100 of the *NEC*. Basically, these are flammable liquids having a flash point below 38°C (100°F) or flammable or combustible liquids raised above their flash points. (See the definition in the *NEC* for more details. See also Sections 3-1.1.2 and 3-1.2 of this book.)

2-4.3.2.2 Division 2

The *National Electrical Code* defines Class I, Division 2 locations as follows:

1. Those locations in which volatile flammable liquids or flammable gases are handled, processed, or used, but in which the liquids, vapors, or gases will normally be confined within closed containers or closed systems from which they can escape only in case of accidental rupture or breakdown of such containers or systems, or in case of abnormal operation of equipment
2. Those locations in which ignitible concentrations of gases or vapors are normally prevented by positive mechanical ventilation and which might become hazardous through failure or abnormal operation of the ventilating equipment
3. Those locations that are adjacent to a Class I, Division 1 location and to which ignitible concentrations of gases or vapors might occasionally be communicated, unless communication is prevented by adequate positive-pressure ventilation from a source of clean air and effective safeguards against ventilation failure are provided

The *National Electrical Code* definition is followed by a fine-print note (explanatory material) giving information on typical locations that are classified as Class I, Division 2.

2-4.3.2.3 Division 3

Some individual companies have arbitrarily and for their own purposes established a Division 3 location. This location usually extends beyond a Division 2 location, and the only basic requirement is for the use of enclosed equipment.

2-4.3.3 Zones

Article 505 of the *National Electrical Code* defines the various zones. Basically, Zone 0 is the most hazardous part of Division 1 as defined in Article 500, Zone 1 is all the rest of Division 1, and Zone 2 is the same as Division 2. The relationship is shown in Figure 2-6.

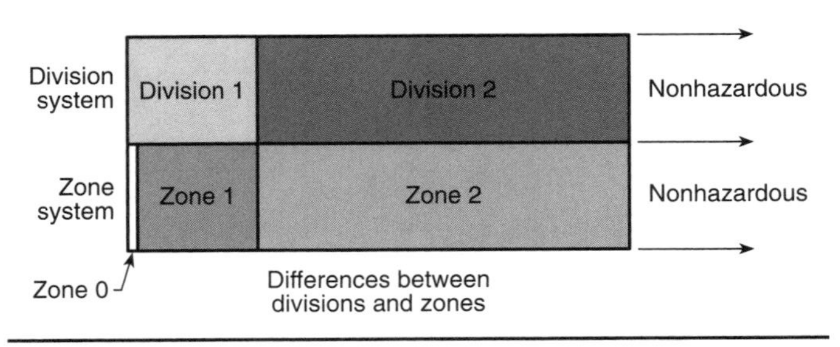

Figure 2-6. *Differences between divisions and zones. (Courtesy of HEP–Killark.)*

2-4.3.3.1 Zone 0

The *National Electrical Code* defines Zone 0 as a location in which ignitible concentrations of gases or vapors are present either (1) continuously or (2) for long periods of time.

This definition is followed by fine-print notes referencing several standards that will help in understanding and classifying areas under the Zone system of area classification as well as information on typical Zone 0 locations.

2-4.3.3.2 Zone 1

Zone 1 locations are defined as follows:

1. Those locations in which ignitible concentrations of flammable gases or vapors are likely to exist under normal operating conditions
2. Those locations in which ignitible concentrations may exist frequently because of repair or maintenance operation or because of leakage
3. Those locations in which equipment is operated or processes carried on, of such nature that equipment breakdown or faulty operations could result in the release of ignitible concentrations and also cause simultaneous failure of electrical equipment in a mode to cause the electrical equipment to become a source of ignition
4. Those locations adjacent to a Zone 0 location from which ignitible concentrations could be communicated, unless communication is prevented by adequate positive-pressure ventilation from a source of clean air and effective safeguards against ventilation failure are provided

Although the definition might appear to include those locations classified as Zone 0, it does not. The key difference is that in Zone 0 the ignitible concentrations are *present continuously,* whereas in Zone 1 they are *likely to be present* under normal operating conditions.

As with the other definitions of divisions and zones, the *National Electrical Code* is followed by explanatory material giving examples of Zone 1 locations.

2-4.3.3.3 Zone 2

The only difference between the definitions of Division 2 and Zone 2 is that the definition of Zone 2 includes those locations in which ignitible concentrations of gases or vapors are not likely to occur in normal operation, and if they do occur, they will exist only for a short period of time.

2-4.3.4 How Divisions and Zones Are Determined

Determining the extent of a hazardous location and the proper classification into divisions or zones can be very simple if the location is common enough that considerable experience has been built up, permitting specific dimensional information in published standards. It can be very difficult, however, in other situations. For more details, see Section 3-1.1.2.

2-5 Class II Locations

2-5.1 *National Electrical Code* Definition

The *National Electrical Code* defines Class II locations as those that are hazardous because of the presence of combustible dust. The intent is to prevent explosions and fires because of the presence of the dust.

Combustible dust is present in small quantities in almost all occupancies. Most, if not all, organic dusts are combustible, so that the dusts in offices from paper handling and from clothing and carpeting are technically capable of creating an explosion. In such occupancies, however, the quantity of dust present is insufficient to cause an explosion, even if it were all to be thrown into suspension in the air at one time. As indicated in Section 2-1.2, a minimum concentration of dust is necessary before a flame will propagate away from the point of ignition.

It is therefore only those locations where large quantities of dusts are or may be present that are classified as Class II locations. For more information on classification of Class II areas, see Section 3-2.1.

Combustible dust is defined in NFPA 499 as any finely divided solid material 420 microns or less in diameter (that is, material passing through a U.S. No. 40 standard sieve) that presents a fire or explosion hazard when dispersed and that ignites in air or other gaseous oxidizer.

2-5.2 Groups

2-5.2.1 Why They Exist

Class II locations are divided into three groups identifying the specific materials involved. However, unlike Class I materials, the reasons for the division into groups are not based on flame paths and explosion pressures. They are based on the electrical resistivity of the dusts, commonly encountered dust particle sizes, and the thermal blanketing effect of dust layers on electrical equipment. Originally, they were also based on layer ignition temperatures of the various dusts, but this concept has been greatly diluted.

2-5.2.2 What They Are

The *National Electrical Code* divides Class II locations into three groups, Groups E, F, and G. The arrangement into groups first appeared in the 1937 edition of the *NEC*, with the groups classified as follows:

Group E: Atmospheres containing metal dust

Group F: Atmospheres containing carbon black, coal, or coke dust

Group G: Atmospheres containing grain dust

These groups remained essentially the same through the 1981 edition of the *NEC,* except for the addition of resistivity limits and chemical and plastics dusts. In 1981 the groups were identified as follows:

Group E: Atmospheres containing metal dust, including aluminum, magnesium, and their commercial alloys, and other metals of similarly hazardous characteristics having a resistivity of 10^2 ohm-centimeter or less

Group F: Atmospheres containing carbon black, charcoal, coal, or coke dusts that have more than 8 percent total volatile material (carbon black per ASTM D 1620, charcoal, coal, and coke dusts per ASTM D 271) or atmospheres containing these dusts sensitized by other materials so that they present an explosion hazard, and having resistivity greater than 10^2 ohm-centimeter but equal to or less than 10^8 ohm-centimeter

Group G: Atmospheres containing flour, starch, grain, or combustible plastics or chemical dusts having a resistivity greater than 10^8 ohm-centimeter

In the 1984 edition of the *NEC* only, Group F was deleted and the following definitions were used:

Group E: Atmospheres containing combustible metal dusts regardless of resistivity or other combustible dusts of similarly hazardous characteristics having resistivity of less than 10^5 ohm-centimeter

Group G: Atmospheres containing combustible dusts having a resistivity of 10^5 ohm-centimeter or greater

In the 1987 edition of the *NEC*, Group F was reinstated for the reasons noted below with the same definition found in the 1981 edition. The definitions of Groups E and G remained as they were in the 1984 edition. In the 1990 edition, the definitions of Groups E and G were revised to be the same as in the 1981 edition. In the 1993 edition, specific resistivities for each group were deleted, because of difficulty in finding a reproducible test method for measuring resistivity. The following definitions appeared in the 1993 edition:

Group E: Atmospheres containing metal dusts, including aluminum, magnesium, and their commercial alloys, and other combustible dusts whose particle size, abrasiveness, and conductivity present similar hazards in the use of electrical equipment

Group F: Atmospheres containing combustible carbonaceous dusts, including carbon black, charcoal, coal, or coke dusts that have more than 8 percent total entrapped volatiles, or dusts that have been sensitized by other materials so that they present an explosion hazard

Group G: Atmospheres containing combustible dusts not included in Groups E or F, including flour, grain, wood, plastic, and chemicals

In the 1996 edition of the *NEC,* coke dust was deleted from Group F.

The thought behind eliminating Group F and dividing Groups E and G by the electrical resistivity of 10^5 ohm-centimeter was that there did not appear to be any dusts with resistivities in the 10^5 ohm-centimeter range. In addition, there did not appear to be any reason either from the particle size or the thermal blanketing effect standpoints for continuing Group F. It was understood that coal dusts would be classified Group G, because they were all thought to have resistivities greater than 10^5 ohm-centimeter.

However, it was found that a few coal dusts from the western part of the U.S. had resistivities slightly less than 10^5 ohm-centimeter under some test conditions. When Group F was dropped in the 1984 edition of the *NEC*, such dusts fell into Group E.

Because coal users such as electric utilities could not guarantee that they would not use some of these coals, demand increased for motors and other equipment approved for Group E. Such equipment was scarce and difficult to build because of the particle size of the dust used in the testing of equipment for Group E location. In addition, equipment suitable for Group G could fail a dust penetration test using dusts considered representative of Group E locations.

Furthermore, the established temperature limits on the external surfaces of equipment were higher for Group E location equipment than was considered safe for coal dusts, and the temperature limits for Group G location equipment were lower than was considered the minimum for Group F location equipment. This presented the possibility of redesigning equipment that had for many years been found acceptable for coal-handling facilities and that had been approved for Group F locations only.

As a result of these practical problems, the National Electrical Code Committee reinstated Group F in the 1987 edition of the *Code.*

2-5.2.3 How the Groups Are Determined

There are two basic criteria for classifying a dust into one of the three groups. The type of material involved is the most important factor. The other is electrical resistivity.

Traditionally, metal dusts have been classified Group E because they present two special problems. First, they are mechanically abrasive. If they enter mechanical bearings the resultant heat buildup can cause ignition of a dust layer on the outside of the bearing, whether this bearing is part of a piece of electrical equipment or simply part of a mechanical system, such as a conveyor belt. Second, metal dusts are generally electrically conductive. If they enter enclosures containing uninsulated live parts, they can cause the breakdown of electrical insulation between these live parts.

Grain dusts and dusts from other foods, most chemical dusts, and plastic dusts, on the other hand, are less abrasive and less likely to cause electrical insulation breakdowns.

The dusts from carbonaceous materials such as coal are somewhere in between, as far as the likelihood of electrical insulation breakdown is concerned. Experience in underground coal mines has shown that low-voltage (600 V and below) systems present no problems, even if coal dust accumulates on electrical insulation between live parts. Above 600 V, however, problems have occurred.

2-5.3 The Role of Ignition Temperature

2-5.3.1 What It Is

In Class II locations, dust may be either suspended in the air in a cloud or in a layer on electrical (or other) equipment. If it is suspended in the air in a cloud, it is usually also present in a layer on equipment. The temperature required to cause ignition of a dust layer on a heated surface is almost always lower than it is for a dust cloud in a heated chamber. In those cases in which the cloud ignition temperature recorded has been lower than the layer ignition temperature, there were reasons for this difference, such as a change in state of the dust from a solid to a gas, a slightly different material, and so forth.

Most studies in the United States on the ignition temperature of combustible dusts were conducted by the Bureau of Mines and reported on in a series of reports published in the 1960s. The results are summarized in NMAB 353-2[29] and NMAB 353-4. See Appendix A, Table A-6, for the ignition temperatures of some dusts.

2-5.3.2 How It Is Determined

Layer Ignition Temperature: The ignition temperature of a dust layer is normally determined by some type of hot-plate test method, whereby a known thickness and particle distribution size of the test dust is slowly heated on an electrically heated plate until ignition occurs. Such test methods are described in the various Bureau of Mines reports and in publications NMAB 353-2 and 353-4. The method described in the publications NMAB 353-2 and 353-4 are consistent with the international test method.

Cloud Ignition Temperature: The method used by the Bureau of Mines to determine the ignition temperature of a dust cloud consisted essentially of a heated furnace into which a cloud of dust of known concentration was introduced. The method is described in detail in the various Bureau of Mines reports, such as U.S. Bureau of Mines RI 5624.[30]

2-5.3.3 Where to Go for Information on Unknowns

NFPA Material: NFPA 499 provides information on the minimum cloud or layer ignition temperature of a great many dusts (this material was originally in NFPA 497M). This information is based on the publications NMAB 353-2 and 353-4, which in turn are based on the Bureau of Mines reports RI 5753[31], RI 5971[32], RI 6516[33], RI 6597[34], RI 7132[35], and RI 7208[36].

2-5.4 Divisions

2-5.4.1 Why They Exist

The subdivision of Class II locations into divisions represents, in effect, the likelihood of a dust explosion if ignition sources are not carefully controlled. A very common sequence of events in a dust explosion would begin with a dust layer being ignited in some manner, such as by a spark from a welding operation, an overheated bearing, or an unprotected electric light bulb. The resulting (often smoldering) fire creates turbulence, which propels the layers of accumulated dust into the air. This small dust cloud is ignited by the fire, creating a small explosion. The small explosion, in turn, throws more dust into suspension, creating a larger explosion. Dust explosions are often described as a long series of rumbling noises gathering in intensity.

Control of dust layers is extremely important, as dust explosions are commonly started by the ignition of a layer of dust. The separation of Class II locations into divisions is in recognition of this fact.

2-5.4.2 What They Are

Class II, Division 1: The *National Electrical Code* defines Class II, Division 1 locations as follows:

1. Those locations in which combustible dust is in the air under normal operating conditions in quantities sufficient to produce explosive or ignitible mixtures
2. Those locations where mechanical failure or abnormal operation

of machinery or equipment might cause such explosive or ignitible mixtures to be produced and might also provide a source of ignition through simultaneous failure of electric equipment, operation of protective devices, or from other causes

3. Those locations in which combustible dusts of an electrically conductive nature may be present in hazardous quantities

The definition of Class II, Division 1 locations includes those locations in which combustible dusts of an electrically conductive nature may be present in hazardous quantities. The term "hazardous quantities" is not defined, but quantities are generally considered to be hazardous whenever there is sufficient metal dust present to form a visible film over equipment, thus obscuring the surface color. (See Section 3-2.1.1 for more information.)

To make the intent clearer, the *NEC* also indicates that when Class II, Group E dusts are present in hazardous quantities, there are only Division 1 locations. There are two reasons for this:

1. Electrically conductive dusts can cause electrical equipment breakdowns if they are deposited on electrical insulating materials supporting live parts, thus creating an ignition source that would not have existed had the dust not been electrically conductive.
2. Metal dusts, in particular, are abrasive, and if they enter the bearing housings of rotating electrical machinery, overheating of the bearings and ignition as a result of this overheating can result.

Table 2-5, which was taken from NFPA 499, is provided as a guide.

Class II, Division 2: The National Electrical Code defines Class II, Division 2 locations as those locations where combustible dust is not

Table 2-5 Guide to Classification of Class II Locations by Division

Thickness of Dust Layer on Equipment*	Dust Group	Division
Greater than ⅛ in. (3 mm)	E, F, G	1
⅛ in. (3 mm) or less but surface color not discernible	E	1
⅛ in. (3 mm) or less but surface color not discernible	F, G	2
⅛ in. (3 mm) or less and surface color discernible	E, F, G	Unclassified under dust layer

**Based on buildup of dust level in a 24-hour period on the major portions of the horizontal surfaces.*

normally in the air in quantities sufficient to produce explosive or ignitible mixtures, and dust accumulations are normally insufficient to interfere with the normal operation of electrical equipment or other apparatus, but combustible dust may be in suspension in the air as a result of infrequent malfunctioning of handling or processing equipment and where combustible dust accumulations on, in, or in the vicinity of the electrical equipment may be sufficient to interfere with the safe dissipation of heat from electrical equipment or may be ignitible by abnormal operation or failure of electrical equipment.

This definition is followed by fine-print notes (explanatory material) indicating that the quantity of combustible dust that may be present and the adequacy of dust removal systems are factors that merit consideration in determining the classification and may result in an unclassified (ordinary location) area. There is also a note indicating that where products such as seed are handled in a manner that produces low quantities of dust, the amount of dust deposited may not warrant classification.

Both of these notes were added to the *National Electrical Code* to make it clear that it is not the intent of the *NEC* to classify areas simply because there is or might be some dust present. It is intended that dust removal systems, common in many industries and highly recommended by fire safety experts, be recognized as a method of changing to nonhazardous an area that might otherwise be considered a Class II, Division 2 location. In some industries, such as the seed-handling industry, great care is taken to be sure that the outer covering of the seed is not damaged during handling. This, in turn, reduces the amount of dust generated, even though an area processing the same material for different purposes might well be classified as a Class II location.

2-5.4.3 How They Are Determined

A definitive classification of Class II locations into divisions is much more difficult than classification of Class I locations into divisions, because the classification process does not readily lend itself to mathematical approaches. NFPA 499 provides guidance on area classification in Class II locations. See Section 3-2.1.1 of this book for additional information.

2-6 Class III Locations

2-6.1 *National Electrical Code* Definition

The *National Electrical Code* defines a Class III location as one that is hazardous because of the presence of easily ignitible fibers or flyings, but in which such fibers or flyings are not likely to be in suspension in the air in quantities sufficient to produce ignitible mixtures.

2-6.2 Materials

There are no group subdivisions in Class III hazardous locations. The materials normally considered within Class III locations include the easily ignitible fibers and flyings from rayon, cotton (including cotton linters and cotton waste), sisal or henequen, istle, jute, hemp, tow, cocoa fiber, oakum, baled waste kapok, Spanish moss, and excelsior. Sawdust in woodworking plants, where it can accumulate quickly and in large quantities and where it can dry out and present a severe fire hazard, is considered a Class III material. Wood flour, on the other hand, which is from the same basic material, is a Class II material, because it can be in suspension in the air in the same way that finely divided dusts from grain can.

The major hazard from Class III materials is not an explosion hazard, but a fire hazard. Fire travels extremely rapidly through these materials.

A typical example of a Class III material that exists in many homes is the lint that can be found in every clothes dryer. In this case, the amount of material present is insufficient to result in classification of the area as a Class III location. In fabric mills, however, essentially the same material is present in large quantities. Collections of the material on roof trusses, ledges, and other horizontal surfaces can cause a fire to travel with extreme rapidity throughout a very large building.

2-6.3 Divisions

Although Class III locations are not grouped based on the type of material present, they are divided into two divisions based on the likelihood of the material being present in dangerous quantities.

2-6.3.1 Why They Exist

Divisions exist in Class III locations for the same reason they exist in Class I and Class II locations: because the degree of protection necessary is greater in those locations where the material is more likely to be present than it is in locations where it is less likely to be present.

2-6.3.2 What They Are

A Class III, Division 1 location is a location in which the easily ignitible fibers or materials producing combustible flyings are handled, manufactured, or used.

A Class III, Division 2 location is a location where easily ignitible fibers are stored or handled except in the process of manufacture (where it is a Division 1 location).

2-6.3.3 How They Are Determined

Unlike the difficulty in determining the extent of a Class I, Division 1 or Division 2 location, or a Class II, Division 1 or Division 2 location, it is relatively simple to determine the extent of Division 1 and Division 2 locations in Class III locations. It depends on the use of the area rather than the amount of material present or likely to be present. For additional information, see Section 3-3.1.

2-7 Grounding and Bonding

Grounding and bonding is necessary for electrical safety in nonhazardous locations as well as in hazardous locations. In hazardous locations it is vital to have good grounding and bonding to prevent an explosion.

Explosionproof equipment is designed to withstand an internal explosion; dust-ignitionproof equipment is designed to prevent the entrance of dust into the enclosure. In all cases the enclosures are relatively rugged, and the established construction requirements normally specify a minimum wall thickness for metal enclosures. Should there be an arcing fault to a metal enclosure, it is only a matter of time until the external surface of the enclosure at the point of the fault reaches a temperature that can cause ignition of flammable gases or accumulations of combustible dust. A good low-impedance path back to the panelboard or switchboard in which the circuit originates is essential to assure proper operation of overcurrent protective devices, such as fuses and circuit breakers. Under fault conditions it is necessary that this overcurrent device be caused to operate as quickly as possible to prevent a hot spot on the enclosure or even burn through, and ignition of flammable atmospheres or layers of combustible dust or easily ignitible fibers or flyings.

Under fault current conditions when heavy currents are flowing through metal conduit, every connection point in the conduit system is a potential source of sparks and therefore a potential source of ignition. It is most important, therefore, that all threaded joints be wrenchtight to prevent sparking at the threads. If joints are other than of the threaded type, such as locknuts and bushings or double locknuts and bushings at boxes, cabinets, and panelboards, it is essential that a bonding jumper be used around such joints in the grounding path to prevent sparking and to assure a low-impedance path.

The *National Electrical Code* indicates that the locknut-bushing and double-locknut types of contacts are not to be depended upon for bonding purposes in hazardous locations. It also requires this bonding for all intervening raceways, fittings, boxes, enclosures, and so forth between the hazardous location and the point of grounding for service equipment, separately derived system, or building supply ground, even if there are intervening overcurrent devices. This is intended to assure operation of the entire protective system.

In addition to the possibility of arcs and sparks or overheated parts as a result of faults, grounding is necessary to prevent the buildup of static electrical charges. In this case, however, a low-impedance path is not essential. For additional information, see Section 5-1.

In intrinsically safe systems, particularly those using diode barriers or the equivalent, an extremely low-impedance path to a common ground point is needed to assure proper operation of the safety devices provided.

2-8 Ignition Sources

In order for combustion (fire or explosion) to occur, three elements must be present: fuel, heat (that is, an ignition source), and oxygen or air. The contribution each of these elements makes in the production of a fire or explosion is depicted in Figure 2-7 as the fire triangle.

In the hazardous locations being considered here, the fuel is a flammable gas, a vapor from a flammable or combustible liquid, combustible dust, or easily ignitible fibers or flyings. One of these fuels must then be mixed in air or oxygen in proportions that will produce an explosive mixture. Finally, an ignition source of sufficient energy must be present to ignite the mixture.

Obviously, if the fuel is not present, the area is not classified as a hazardous location. Assuming, therefore, that the fuel is present and air or oxygen is also present, the most common method of preventing a fire or explosion is to eliminate the heat or ignition source side of the fire triangle. These ignition sources are addressed in the following paragraphs.

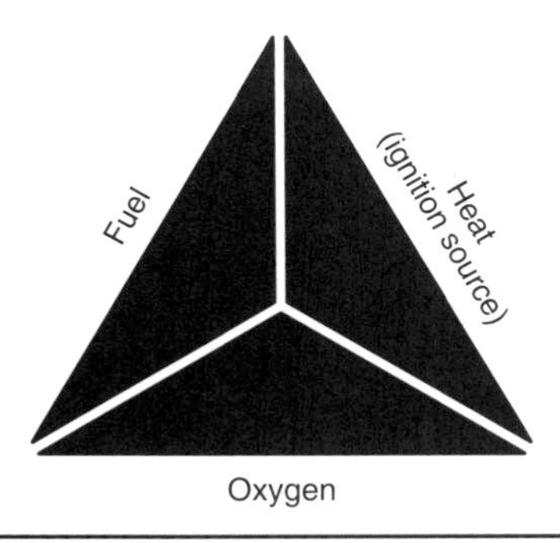

Figure 2-7. *Fire triangle.*

2-8.1 Arcs and Sparks

A certain finite quantity of energy is needed to ignite flammable and combustible materials. The energy necessary depends upon the particular material involved, the concentration of this material in the air or with an oxidizing agent, the electrode shapes and materials, and the time period over which the energy release occurs. Only a very small amount of energy is needed to cause ignition of common flammable gases, as shown in Table 2-6. (See also Table 5-1 in Chapter 5.)

For combustible dusts the minimum ignition energies are many times as great as they are for gases, as shown in Table 2-7.

Determinations of minimum ignition energy are normally made using capacitors charged to several thousand volts so that the energy is released into the flammable gas-air or dust-air mixture almost instantaneously and with a minimum of loss due to heating of the electrodes. The minimum ignition energy for a material is determined by adjusting the gap between the electrodes and the flammable mixture until the most easily ignited concentration of mixture and the minimum ignition energy for this mixture have been determined. In actual practice, the energy required to ignite a flammable mixture will always be higher than this

Table 2-6 Minimum Ignition Energy, Gases

Flammable Material	Minimum Ignition Energy (Gases) (millijoules)
Acetylene	0.017
Hydrogen	0.017
Ethylene	0.08
Methane	0.3

Note: from Bureau of Mines RI 5671[37].

Table 2-7 Minimum Ignition Energy, Dusts

Combustible Material	Minimum Ignition Energy (Dusts) (millijoules)
Aluminum	15
Magnesium	40
Grain	240
Soft coal	30
Hard coal	60
Sulfur	15
Phenol formaldehyde	15

Note: from NMAB 353-4.

minimum value, because all of the ideal conditions (most easily ignited mixture, no loss of energy in heating of electrodes, and maximum energy release in minimum time) are not achieved. These conditions can be approached, however, with static electrical sparks.

For a more detailed study of ignition by arcs and sparks, the authors recommend *Electrical Instruments in Hazardous Locations*[38].

2-8.1.1 Power Arcs from Electrical Equipment

Electrical equipment such as motor controllers, circuit breakers, switches, and relays is needed and commonly used in hazardous locations. Except for some signaling, communications, and data processing circuits, where the energy released at contacts can be maintained below the critical ignition energy, the arc at the contacts of such switching equipment is ignition capable. The contacts of such equipment must therefore be enclosed in such a manner that unwanted explosions are prevented. (See Chapter 4 for equipment protection systems.)

In addition to normally anticipated arcs at contacts of switching equipment, arcs can occur as a result of insulation breakdown, either between live parts at different polarities or between live parts and grounded metal.

Arcs can also occur when a wire breaks or fuses open. For example, if an incandescent lamp bulb is broken without breaking the lamp filament, the filament will very quickly burn open when exposed to the oxygen in the air. So too will an electrical conductor when subjected to a high level of current, such as a ground fault or short circuit, unless both the conductor and its overcurrent protection are properly sized based on the available short-circuit current in the circuit. Fuses are a prime example of a wire (the fusible element) burning open under overcurrent conditions.

Arcs also occur at slip rings and commutators on rotating electrical equipment.

When an electrical conduit is subjected to a high level of fault current, sparks will normally be observed at joints that are loose or corroded. (See Section 2-7 regarding the need for proper grounding and bonding.)

2-8.1.2 Static Electricity (Except Lightning)

Static electricity is a common source of ignition of flammable atmospheres, and is treated separately in Section 5-1.

2-8.1.3 Lightning

Lightning, a very severe and high-energy form of static electricity, is another source of ignition. Although it cannot be completely controlled, the likelihood of lightning igniting a flammable atmosphere can be reduced. (See Section 5-2.)

2-8.1.4 Mechanical Sparks

Mechanical sparks created by two objects striking each other are also a common cause of explosions. Some grain elevator dust explosions have been attributed to elevator buckets striking the sides of the shaft or other parts of the elevator mechanism.

Very little work has been done in determining the ignition energy from mechanical sparks. That such sparks can cause ignition is quite evident, however, if one simply considers that flint and steel have been used for centuries for starting fires.

Mechanical sparks can occur wherever two surfaces strike each other, either under normal conditions (such as in grinding operations where showers of sparks are produced) or under abnormal conditions where a moving part accidentally strikes another moving part or a stationary part.

Use of nonsparking metals (such as bronze and some aluminum alloys) or nonmetallic materials can reduce the likelihood of creating mechanical sparks. When the entire part cannot be made of a nonsparking material, a heavy jacket of nonsparking material will sometimes produce the same results.

Care can also be exercised in preventing the introduction of foreign materials (for example, scrap iron) in grinding or crushing operations or into the airstream of fans.

2-8.1.5 Light-Alloy Enclosures

Light-alloy (aluminum and magnesium) enclosures, if struck with sufficient force by rusted steel, can produce ignition-capable sparks as a result of a chemical reaction between the materials. For this reason, explosionproof and flameproof enclosures made of magnesium and aluminum are commonly considered unacceptable in underground mining operations and in aboveground industrial applications in some countries. In the United States, aluminum explosionproof enclosures have been used extensively in industrial applications, with no history of problems.

2-8.2 Open Flames

Although it is obvious that open flames can be a source of ignition, they are sometimes used to prevent transmission of flammable gases to other locations. For example, an exhaust stack that is known to emit flammable gases can be fitted with a continuous flame to ignite flammable gases at the exit from the stack so that they will be burned off. The flares in oil refineries are a typical example of the use of open flames to prevent transmission of flammable gases.

Included with open flames are glowing combustibles, such as from smoking materials.

2-8.2.1 Welding and Cutting Operations

Improper welding and cutting operations (often known as ''hot work'') are also a common cause of explosions in hazardous locations. The open flame or exposed arc is an obvious ignition source. Sometimes overlooked, however, is the ignition source created by the hot welds after the welding operation is completed, as well as the shower of hot metal resulting from cutting operations.

One major grain elevator explosion was traced to hot work. The report of the investigation indicated that such work was being done on the spout of a grain discharge chute. The operator stopped the flow of grain while the welding was in progress and asked the welding crew chief to advise him as soon as the operation had been completed. This was done, but an explosion occurred as soon as the flow of grain commenced. Ignition was from the still-hot weld[39].

In another case, a crew was repairing the inner tank lining of a liquefied natural gas tank. The tank had been purged with air before the crew entered and started operations. The tank was not repurged, however, during the lunch break, and when welding operations started again, an explosion occurred. The insulation between the inner and outer walls of the tank had become permeated with natural gas as a result of the leak in the inner tank that was being repaired, and enough accumulated during the lunch break to result in an explosive mixture within the tank. Natural gas in liquefied natural gas tanks is normally stored without an odorant so that its presence was not detected by the workmen when they re-entered the tank after lunch[40].

Control of work practices is the only effective method for prevention of ignition by welding and cutting operations.

2-8.2.2 Ovens and Furnaces

Because of the high temperatures involved, ovens and furnaces are an obvious source of ignition. In addition, many ovens and furnaces are fired by fossil fuels, so that there is also the likelihood of a flammable atmosphere being present. However, if properly installed, maintained, and operated, ovens and furnaces are not a source of uncontrolled ignition — either for the fossil fuel being used or for flammable vapors given off by materials being dried, such as newly painted pieces in the drying oven at the end of a spray finishing line. The key to this safety is maintaining a continuous flow of air such that the flammable vapors are always below the lower flammable limit.

The most dangerous period for ignition possibility is during startup and shutdown of the oven or furnace, when the airflow may be reduced or cut off, permitting a buildup of flammable vapors to within the flammable range.

For more details as to the proper safety precautions, see the appropriate codes and standards, such as NFPA 86, *Standard for Ovens and Furnaces*[41], NFPA 86C, *Standard for Industrial Furnaces Using a Special Processing Atmosphere*[42], NFPA 86D, *Standard for Industrial Furnaces Using Vacuum as an Atmosphere*[43], and NFPA 91, *Standard for Exhaust Systems for Air Conveying of Materials*[44].

2-8.2.3 Gas Turbines and Combustion Engines

Because of the large volume of air passing through a gas turbine and a combustion engine, the turbine and engine are not normally a source of ignition. For additional information on precautions for combustion engines and gas turbines, see NFPA 37, *Standard for the Installation and Use of Stationary Combustion Engines and Gas Turbines*[45].

Rooms or other locations for gas turbines and combustion engines are not classified as hazardous locations solely by reason of the turbine or engine fuel. This does not mean that external electrical parts of the gas turbine do not need some protection. NFPA 37 requires a degree of protection for the ignition system.

2-8.3 Hot Surfaces

Hot surfaces are a potential source of ignition. For additional information, see Section 2-4.2.4 on the role of ignition temperature in Class I locations and Section 2-5.3 on the role of ignition temperature in Class II locations. In general, if the temperature of a surface exceeds the ignition temperature of the material involved, it can be considered a potential source of ignition. This consideration must be tempered with other factors. If the flammable material is moving rapidly across the heated surface, such as a flammable gas-air mixture in a turbulent condition or in a stream of mixture flowing across the surface, the mixture will not reach the same temperature as the heated surface.

In addition, if the heated surface is small in comparison to the area in which it is located, the heated surface will not be capable under ordinary conditions of heating the mixture to the temperature of the hot surface. This is typical of electrical equipment installed in a hazardous location, such as lighting fixtures and motors. An extreme example of a small surface is a hot wire.

The general *National Electrical Code* requirement preventing use of equipment having an operating temperature greater than the ignition temperature of the particular flammable material involved has a built-in safety factor. This is one of the reasons the *NEC* permits the operating temperature of the equipment to be at (but not above) the ignition temperature.

In those cases where the ignition temperature or temperature range is marked on the equipment and the equipment is listed or certified, the equipment has been tested and the operating temperature is known. In some cases, the *National Electrical Code* indicates that the operating temperature of the equipment is not to exceed 80 percent of the ignition temperature in degrees Celsius. This originally was intended to add another safety factor into the requirements for equipment that was not specifically listed or certified for use in the environment. For example, for many years electric lighting fixtures in Class I, Division 2 hazardous locations were not specifically required to be approved for those locations, and ordinary-location fixtures with the lamps enclosed were considered acceptable. The actual temperatures of the lamp surfaces on these fixtures were not known; they were estimated, often based on average temperature information published by lamp manufacturers.

2-8.4 Other Ignition Sources

There are other potential sources of ignition that must be taken into consideration. Included are high-frequency electromagnetic waves, such as those from radio, television, and radar installations; optical radiation, such as from laser sources; ionizing radiation, such as from radioactive thickness gauges; ultrasonic waves, such as from density-measuring equipment; sudden compression, such as from failure of a vacuum system; chemical reactions, such as spontaneous ignition; and stray currents from corrosion processes, ground currents, and so forth.

Bibliography

[1]NFPA 325, *Guide to Fire Hazard Properties of Flammable Liquids, Gases, and Volatile Solids,* National Fire Protection Association, Quincy, MA, 1994.

[2]NFPA 70, *National Electrical Code®,* National Fire Protection Association, Quincy, MA, 1996.

[3]NFPA 30, *Flammable and Combustible Liquids Code,* National Fire Protection Association, Quincy, MA, 1996.

[4]ASTM D 56, *Method of Test for Flashpoint by Tag Closed Tester,* American Society for Testing and Materials, Philadelphia, PA, 1987.

[5]ASTM D 93, *Methods of Test for Flashpoint by Pensky-Martens Closed Tester,* American Society for Testing and Materials, Philadelphia, PA, 1990.

[6]ASTM D 3278, *Standard Test Methods for Flashpoint of Liquids by Setaflash — Closed-Cup Apparatus (Paints/Enamels/Lacquer/Varnishes),* American Society for Testing and Materials, Philadelphia, PA, 1995.

[7]ASTM D 3828, *Standard Test Method for Flashpoint of Liquids by Small Closed-Cup Tester,* American Society for Testing and Materials, Philadelphia, PA, 1993.

[8]ASTM D 1310, *Standard Test Method for Flashpoint and Fire Points of Liquids by Tag Open-Cup Apparatus,* American Society for Testing and Materials, Philadelphia, PA, 1986.

[9]NFPA 53, *Guide on Fire Hazards in Oxygen-Enriched Atmospheres,* National Fire Protection Association, Quincy, MA, 1994.

[10]ASTM E 681, *Standard Test Methods for Concentration Limits of Flammability of Chemicals,* American Society for Testing and Materials, Philadelphia, PA, 1994.

[11]NMAB 353-4, *Classification of Dusts Relative to Electrical Equipment in Class II Hazardous Locations,* National Academy Press, Washington, DC, 1982. (Available from National Technical Information Service, Springfield, VA 22151.)

[12]Bulletin of Research No. 58, *An Investigation of 15 Flammable Gases or Vapors with Respect to Explosion-Proof Electrical Equipment,* Underwriters Laboratories Inc., Northbrook, IL, 1969.

[13]NFPA 497M, *Manual for Classification of Gases, Vapors, and Dusts for Electrical Equipment in Hazardous (Classified) Locations,* National Fire Protection Association, Quincy, MA, 1991. (Discontinued.)

[14]NFPA 497, *Recommended Practice for the Classification of Flammable Liquids, Gases, or Vapors and of Hazardous (Classified) Locations for Electrical Installations in Chemical Process Areas,* National Fire Protection Association, Quincy, MA, 1997.

[15]NFPA 499, *Recommended Practice for the Classification of Combustible Dusts and of Hazardous (Classified) Locations for Electrical Installations in Chemical Process Areas,* National Fire Protection Association, Quincy, MA, 1997.

[16]Bulletin of Research No. 58A, *An Investigation of Additional Flammable Gases or Vapors with Respect to Explosion-Proof Electrical Equipment,* Underwriters Laboratories Inc., Northbrook, IL, June 1976.

[17]Bulletin of Research No. 58B, *An Investigation of Additional Flammable Gases or Vapors with Respect to Explosion-Proof Electrical Equipment,* Underwriters Laboratories Inc., Northbrook, IL, July 1997.

[18]NMAB 353-6, *Rationale for Classification of Combustible Gases or Vapors with Respect to Explosion-proof Electrical Equipment,* National Academy Press, Washington, DC, 1982. (Available from National Technical Information Service, Springfield, VA 22151.)

[19]Strehlow, R.A., Nicholls, J.A., Magison, E.C., and Schram, P.J., "An Investigation of the Maximum Experimental Safe Gap Anomaly," *Journal of Hazardous Materials,* Vol. 3, 1979, pp. 1–15.

[20]Brown, G.K., Dairty, E.D., and D'Aoust, A., "The Variation of Maximum Experimental Safe Gap with Secondary Explosion Chamber Relief for Ether-Air Mixtures," Division Report FRC 71/53, CEAL 242, Department of Energy, Mines and Resources, Ottawa, Canada, August 1971.

[21]NMAB-447, *Proceedings of the International Symposium on the Explosion Hazard Classification of Vapors, Gases, and Dusts,* National Academy Press, Washington, DC, 1987. (Available from National Technical Information Service, Springfield, VA 22151.)

[22]ASTM D 286, *Standard Method of Test for Autogenous Ignition Temperatures of Petroleum Products,* American Society for Testing and Materials, Philadelphia, PA, 1930. (Discontinued — replaced by ASTM D 2155.)

[23]ASTM D 2155, *Standard Method of Test Autoignition Temperature of Liquid Petroleum Products,* American Society for Testing and Materials, Philadelphia, PA, 1966. (Discontinued — replaced by ASTM E 659.)

[24]ASTM E 659, *Standard Test Method for Autoignition Temperature of Liquid Chemicals,* American Society for Testing and Materials, Philadelphia, PA, 1978 (revised 1984).

[25]NFPA 49, *Hazardous Chemicals Data,* National Fire Protection Association, Quincy, MA, 1994.

[26]NFPA 704, *Standard System for the Identification of the Hazards of Materials for Emergency Response,* National Fire Protection Association, Quincy, MA, 1996.

[27]NMAB 353-5, *Classification of Gases, Liquids, and Volatile Solids Relative to Explosion-Proof Electrical Equipment,* National Academy Press, Washington, DC, 1982. (Available from National Technical Information Service, Springfield, VA 22151.)

[28]IEC Report, Publication 70-10, second edition, *Electrical Apparatus for Explosive Gas Atmospheres, Part 10: Classification of Hazardous Areas,* International Electrotechnical Commission, Geneva, Switzerland, 1986.

[29]NMAB 353-2, *Test Equipment for Use in Determining Classification of Combustible Dusts,* National Academy of Sciences, Washington, DC, 1979. (Available from National Technical Information Service, Springfield, VA 22151.)

[30]RI 5624, *Laboratory Equipment and Test Procedures for Evaluating Explosibility of Dusts,* U.S. Bureau of Mines, Philadelphia, PA. (Available from National Technical Information Service, Springfield, VA 22151.)

[31]RI 5753, Explosibility of Agricultural Dusts, U.S. Bureau of Mines, Philadelphia, PA. (Available from National Technical Information Service, Springfield, VA 22151.)

[32]RI 5971, Explosibility of Dusts Used in Plastics Industry, U.S. Bureau of Mines, Philadelphia, PA. (Available from National Technical Information Service, Springfield, VA 22151.)

[33]RI 6516, *Explosibility of Metal Powders,* U.S. Bureau of Mines, Philadelphia, PA. (Available from National Technical Information Service, Springfield, VA 22151.)

[34]RI 6597, *Explosibility of Carbonaceous Dusts,* U.S. Bureau of Mines, Philadelphia, PA. (Available from National Technical Information Service, Springfield, VA 22151.)

[35]RI 7132, *Dust Explosibility of Chemicals, Drugs, Dyes, and Pesticides,* U.S. Bureau of Mines, Philadelphia, PA. (Available from National Technical Information Service, Springfield, VA 22151.)

[36]RI 7208, *Explosibility of Miscellaneous Dusts,* U.S. Bureau of Mines, Philadelphia, PA. (Available from National Technical Information Service, Springfield, VA 22151.)

[37]RI 5671, *Minimum Ignition-Energy Concept and Its Application to Safety Engineering,* U.S. Bureau of Mines, Philadelphia, PA. (Available from National Technical Information Service, Springfield, VA 22151.)

[38]*Electrical Instruments in Hazardous Locations,* third edition, Instrument Society of America, Research Triangle Park, NC, 1978.

[39]McKinnon, G.P., editor, *Fire Protection Handbook,* 14th edition, National Fire Protection Association, Quincy, MA, 1976.

[40]"Multiple Death Fires, 1973," *Fire Journal,* Vol. 68, No. 3, May 1974, p. 72.

[41]NFPA 86, *Standard for Ovens and Furnaces,* National Fire Protection Association, Quincy, MA, 1995.

[42]NFPA 86C, *Standard for Industrial Furnaces Using a Special Processing Atmosphere,* National Fire Protection Association, Quincy, MA, 1995.

[43]NFPA 86D, *Standard for Industrial Furnaces Using Vacuum as an Atmosphere,* National Fire Protection Association, Quincy, MA, 1995.

[44]NFPA 91, *Standard for Exhaust Systems for Air Conveying of Materials,* National Fire Protection Association, Quincy, MA, 1995.

[45]NFPA 37, *Standard for the Installation and Use of Stationary Combustion Engines and Gas Turbines,* National Fire Protection Association, Quincy, MA, 1994.

CHAPTER 3

Classification of Specific Processes/Occupancies

3-1 Classification of Class I Hazardous Locations

The most difficult part of area classification is applying the defined area classifications to an actual installation. Standards and recommended practices exist for a number of processes and occupancies. However, these documents often are not available to those responsible for the electrical design, installation, or inspection. This chapter will focus on guidelines for the classification of specific processes or occupancies. In each section, installations and problems typical in occupancies of that type will be discussed. The electrical requirements and recommendations in various NFPA codes, standards, and recommended practices have been extracted and are used as the basis to determine the extent of area classification from the source of release of the flammable gas, combustible dust, or combustible flying material.

The 1996 edition of the *National Electrical Code*[1] introduced a second classification system for Class I locations. Article 505 of the *NEC* covers Class I, Zone 0, 1, and 2 locations. At this time only a few of the standards discussed in the chapter have addressed this optional classification system. It can be expected that the committees developing these standards will review the Class I, Zone 0, 1, and 2 requirements, and where necessary, they will modify existing requirements.

Although area classification contemplates some system/process failures, it does not usually encompass situations that may involve catastrophic failure of, or catastrophic discharge from, process vessels, pipelines, tanks, or systems. Failures of this kind may be caused by natural disasters such as earthquakes or other events that result in major structural failures. These failures are so rare that it is impractical to design electrical systems for such circumstances. If design of electrical systems was required to encompass catastrophic failure, then it would be necessary to design nonelectrical systems and other practices to contemplate catastrophic failure. This would place severe constraints on the electrical system as well as other systems, such as heating and ventilation, and on motor vehicles, even though they may be located some distance from the source of catastrophic failure.

As indicated in Chapter 2, the explosion characteristics of air mixtures of gases, vapors, or dusts vary with the specific material involved. For Class I locations, Groups A, B, C, and D, the classification involves determinations of maximum explosion pressure of the gas or vapor, the maximum safe clearance between parts of a flanged joint in an enclosure necessary to cool the hot gases, and the minimum ignition temperature of the atmospheric mixture. For Class II locations, Groups E, F, and G, the classification involves the electrical conductivity of the dust and the blanketing effect of layers of dust on the equipment that may cause overheating. Some metal dusts produce very fine powders that are easily ignitible and that may penetrate shaft openings, which may interfere with the safe operation of rotating machinery. Because of this, the joints of assembly and shaft openings must be designed to prevent entrance of ignitible amounts of dust into the dust-ignitionproof enclosure. It is necessary, therefore, that equipment be approved not only for the class, but also for the specific group of the gas, vapor, or dust that will be present.

Certain chemical atmospheres have characteristics that require safeguards beyond those normally required for hazardous locations. Carbon disulfide and oxygen-enriched atmospheres are typical examples. Carbon disulfide is a chemical with a low ignition temperature [100°C (212°F)] and a small joint clearance to arrest its flame. The size of the joint is smaller than that required for Group A. Therefore, there are very few pieces of equipment designed for it. See Section 2-4.2.3.

Processes involving oxygen-enriched atmospheres may also be considered particularly hazardous. In general, the greater the oxygen concentration, the lower the minimum ignition energy required for ignition and the faster the flame spread rate. Materials that are normally considered noncombustible or fire retardant may burn freely in a pure oxygen atmosphere.

Pyrophoric materials are also considered to be particularly hazardous. These materials will ignite upon contact with ordinary air. Such locations are not normally classified under Article 500 of the *National Electrical Code*. When a pyrophoric material is in an inert atmosphere, it cannot be ignited by any electrical equipment that may be present. When a pyrophoric material is exposed to the atmosphere, the use of special electrical equipment becomes a moot point, since the material will be ignited by the atmosphere. For more information on pyrophoric materials, see Section 2-1.3.

3-1.1 General Guidelines for the Classification of Class I Locations

3-1.1.1 Conditions Necessary for Classification

Article 500 of the *National Electrical Code* defines any location in which a flammable gas or vapor is or may be present in the atmosphere in sufficient concentration to produce an ignitible mixture as a hazardous (classified) location. It is important to note that classification of an area is not necessary unless the gas or vapor may be present in the air in explosive or ignitible quantities under normal or abnormal conditions. See Section 2-1.

3-1.1.2 Factors Affecting the Extent of a Class I Hazardous Location

Normal Conditions: Normal conditions include all of the normal functions of a process as well as normal shutdown conditions. Normal conditions may include the release of gas or vapors in dispensing operations such as gasoline stations, open operations such as dip tanks, spraying of flammable materials such as paint, or the release of gas from a vent on a tank. A routine maintenance or housekeeping procedure can also be considered a normal condition that may require classification. Locations where release can be expected under normal conditions are considered to be Division 1 locations.

Abnormal Conditions: An abnormal condition is one in which failure of process equipment causes a release of gas or vapor. It may also include a situation where a ventilation system fails. Most classification schemes assume adequate ventilation. A ventilation system may also be used to reduce the extent of a hazardous location or to reduce the classification. These locations are known as Class I, Division 2 locations. A Class I, Division 2 location will normally exist in the area immediately adjacent to a Division 1 location. In this area, the normal vapor concentration will be below the lower explosive limit. However, the gas or vapor a short distance away may be in the explosive range. A Division 2 location may exist without an adjacent Division 1 location. A Class I, Division 1 location, however, will always have an adjacent Division 2 location. Division 2 locations also include locations in which an abnormal event occurs. Examples include operation of a pressure relief valve or leakage around a flanged or threaded fitting.

Density of Vapor: The density of the vapor will also have an effect on the extent of the hazardous location. Hydrocarbon vapors and gases are generally heavier than air, while hydrogen and methane are lighter than air. In the absence of barriers (such as walls or enclosures) and disturbing forces (such as air currents), the gas or vapor will disperse.

Heavier-than-air vapors will travel rapidly downward and outward; lighter-than-air vapors will travel upward and outward. If the source of the vapors is a single point, the horizontal area (classified according to the vapor) will be a circle, with the source as the center.

For heavier-than-air vapors released at or near grade level, ignitible mixtures are most likely to be found below grade level; next most likely at grade level; with decreasing likelihood of presence as height above grade level increases. In open areas, away from the immediate point of release, freely drifting gases or vapors from a source near grade have seldom reached ignition sources at elevations more than 6 ft or 8 ft (1.83 m or 2.44 m) above grade. Heavier-than-air vapors tend to collect in below-grade spaces, such as pits. For lighter-than-air gases, the opposite is true. There is little or no hazard at or below grade, but greater hazard above grade.

Ventilation: Ventilation plays a very significant role in the extent of a classified area. In unventilated spaces, a hazardous location would be large. The entire space may be classified as a Division 1 location, because vapors will not be dissipated. (See Figure 3-1.) NFPA 30, *Flammable and Combustible Liquids Code*[2], requires ventilation at the rate of 1 ft^3/minute · ft^2 (0.3 m^3/minute · m^2) of solid floor area in enclosed process areas handling or using any flammable or combustible liquid above its flash point. Such ventilation systems are normally intended to reduce the concentration of flammable gas or vapor in the atmosphere to a level well below the lower explosive limit (LEL); 25 percent of the LEL is typical. Ventilation may cause more gas or vapor to be dispersed in one direction and less in another. However, the area classification schemes in NFPA standards will normally be adequate for these conditions. In fact, a higher ventilation rate will aid in the dilution and dispersion of gases and vapors so that the extent of a hazardous location is greatly reduced, and it may possibly result in reclassification as nonhazardous. In outdoor locations, the prevailing wind cannot be used to reduce the extent of the area classification, because the adequacy and reliability of the wind are unpredictable.

A large, mechanically unventilated building could be a Division 1 location. With ventilation, the Division 1 location may be significantly reduced or even eliminated. Sometimes control rooms in buildings are purged with clean air from a positive mechanical ventilation system. The control room is maintained at a slight pressure above atmospheric to prevent vapors outside of the room from entering. This may negate the need for equipment approved for hazardous locations in the control room. Electrical conduits leading into the room will usually require conduit seals to prevent the entrance of vapors into the control room through the conduits. Additional safeguards such as alarms are necessary to detect failure of the purge system. Complete information on purge

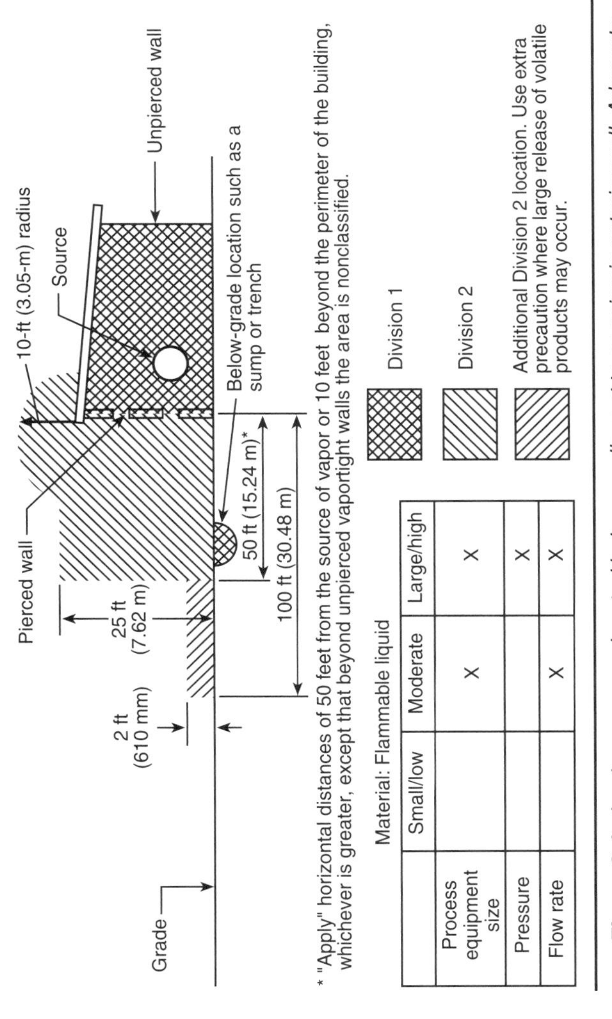

	Small/low	Moderate	Large/high
Process equipment size		X	X
Pressure			X
Flow rate		X	X

Figure 3-1. *Leakage source located indoors, adjacent to opening in exterior wall. Adequate ventilation is not provided. (Source: API Recommended Practice 500A.)*

control rooms can be found in NFPA 496, *Standard for Purged and Pressurized Enclosures for Electrical Equipment*[3]. See also Sections 4-2.1.3 and 4-3.2.4.

Pressure, Flow Rate, and System Capacity: The size of the source will play a significant role in the extent of the classification. This is particularly true for open processes, as the extent of classification extends from the end of the point of release. For a point-of-release source, such as a vent, the location extends radially from the source. For an open process, such as a dip tank, the Division 1 hazardous location includes the entire surface area of the fluid and extends radially above the fluid and for a distance from the inside walls of the tank.

Pressure, flow rate, and system capacity are also very important factors. A release under high pressure [greater than 500 psi (3446 kPa)] can significantly widen the classified area. The same is true of processes where the flow rate of flammable materials is high. Processes without leak detection or some other form of automatic shutoff can permit the spread of flammable materials over a wide area, affecting adjacent areas, drainage systems, and pits. A combustible gas detector can be used to determine the extent of the Division 1 location. (See Figure 3-2.) To determine the extent of the Division 2 location, it is necessary to carefully review the pressure, flow rate, and capacity of the system.

Drainage and Waste Disposal: Drains or curbs are frequently installed around processes using flammable or combustible liquids. Curbs are

Figure 3-2. *A combustible gas detector undergoing calibration tests in a laboratory. (Photo courtesy of Factory Mutual Research Corp.)*

fixed barriers a few inches in height. They are generally made of a noncombustible material such as concrete. Curbs function to contain the flammable liquids, keeping them from spreading far beyond the source of leakage. Drains within the curbed area conduct the fluid away from the process to a sewer system, septic system, or a holding tank. Due to concerns about environmental contamination, holding tanks are now used more frequently than public or private sewer or septic systems.

Some other forms of passive design are frequently used to limit the spread of flammable liquids. Floors may be designed with a pitch toward the process equipment, causing leaked fluids to collect around the process equipment. Vapors given off by the fluids may still travel beyond the process equipment, but their concentration will drop off rapidly with the distance from the pool of liquid. Where no passive protection is provided, a spilled fluid can rapidly spread into adjacent areas. Curbs, drains, and pitched floors are not effective in preventing the spread of flammable gases. The only effective method of preventing accumulations of gas is through the use of ventilation. However, ventilation may not be effective at minimizing a hazard due to a large release of gas.

Storage of Waste Materials: A number of standards address the storage, dispensing, and use of flammable and combustible materials. An issue often overlooked is storage of waste materials prior to disposal. Drums containing waste or contaminated flammable liquids may not be in good condition. Often, the drums are corroded or dented. In some instances, drums are not tightly closed. Such storage may be particularly hazardous. Rooms in which hazardous liquid waste material is stored should be well cut off from the main parts of the plant. Adequate ventilation should be provided to prevent a hazardous accumulation of vapors. The storage room should be classified according to the worst-case material that will be stored in the room. Storage rooms may require classification as Class I, Division 2 locations. Where storage includes drums that are not properly closed, it may be necessary to classify the room as a Class I, Division 1 location. Curbs or a pitched floor should be provided to prevent the spread of a spill into a main plant area. This will eliminate the need to use equipment suitable for hazardous locations in a plant area that would otherwise not require this equipment.

Combustible Gas Analysis: Combustible gas detectors are designed to detect the presence of combustible gases or vapors at some quantity below the lower explosive limit of the gas or vapor. The detectors may be portable or fixed. Fixed instruments are usually provided to sound an alarm, turn on ventilation systems, or shut down processing equipment when the gas concentration exceeds a preset value. Portable instruments are usually used for leak detection or to determine the extent of a vapor concentration.

The most common type of detector is the catalytic detector. These detectors consist of a chamber in which the gas sample is burned on the surface of a heated platinum element. The resistance of the platinum element is measured by a Wheatstone bridge circuit. When the gas is burned, the high temperature associated with combustible gases or vapors increases the resistance of the wire, causing the detector's alarm to sound. The detector can also be designed to provide an analog output, which may provide a digital display of the gas concentration. A more recent development is a type of instrument using solid-state sensors (semiconductors). Generally, these sensors are less specific to combustible gases than the catalytic hot wire filament. However, this type of instrument can be useful as a "go/no go" type of detector.

In operation, gas diffuses through a porous metal disc and comes into contact with the heated surface of a silicon chip. As gas concentrations increase, the resultant chemical reaction reduces the semiconductor resistance logarithmically, producing an output voltage that can be linearized by the instrument's microelectronic circuits.

Gas detectors have different sensitivities to different materials. In order to detect the material in question, the detector must be calibrated for that material. Manufacturers often supply calibration kits that include cylinders of flammable test gases with a known concentration in air or some other inert gas. (See Figure 3-3.)

Other instruments, some generally similar in appearance to the ordinary combustible gas detector, depend for their operation upon properties of a gas or vapor other than combustibility, such as refractive index, density, diffusion, or thermal conductivity of a gas. Calibrated for a specific gas, these instruments can be designed to give readings on high concentrations up to pure gas. The instruments are not selective, however, and will respond to almost any gas or vapor, regardless of combustibility. In some instruments, the catalytic combustion principle and the thermal conductivity principle are combined.

Portable gas detectors are normally used for a number of different gases, whereas a fixed detector may be calibrated for one specific gas and fastened in place. Prior to using a portable instrument, the user should ascertain that the gas detector is not only capable of detecting the gas, but that it is also suitable for the particular hazardous location contemplated. A detector that is not suitable for the location could create a very serious hazard. Listed combustible gas analyzers have a label that indicates the class, division, and groups for which the instrument is suitable. The group of the gas in question can be found in NFPA 497, *Recommended Practice for the Classification of Flammable Liquids, Gases, or Vapors and of Hazardous (Classified) Locations for Electrical Installations in Chemical Process Areas*[4].

Combustible gas analyzers can be a useful tool in determining the extent of a Division 1 location. They may not be as useful for determining the extent of a Division 2 location, because the Division 2 location

Figure 3-3. *Gas detector manufacturers often supply calibration kits that consist of a cylinder of test gas diluted to the desired set point of the detector.*

would have a hazardous concentration of vapors only under abnormal circumstances. Engineering judgment is necessary to determine the limits of the Class I, Division 2 location.

3-1.1.3 Equipment — General Requirements

Articles 500 and 501 of the *National Electrical Code* provide the basic parameters for equipment used in Class I hazardous locations. Article 500 contains definitions, temperature restrictions, and marking requirements. The principle installation requirements can be found in Article 501. Special requirements for specific hazardous locations can be found in Articles 511 through 517.

Equipment for Division 1 locations is designed with the consideration that the location will be continuously or very frequently hazardous. It is therefore necessary to consider normal operation as well as the abnormal operation of the equipment, since it is probable that the atmosphere will be hazardous during an equipment breakdown. The equipment for Division 1 locations must be designed so that even under electrical fault conditions, an explosion cannot take place outside the enclosure. Such equipment will usually include intrinsically safe equipment, explosionproof equipment, or Type X or Type Y purged equipment. (See Chapter 4 for information on each protection method.) Article 501 is written generally with the contemplation of an explosionproof design; however, intrinsically safe or purged equipment may often be substituted in lieu of the explosionproof design.

Division 2 equipment is designed so that it will not ignite a flammable atmosphere under normal operating conditions. It is not designed for a fault condition during exposure to hazardous gases or vapors.

Division 2 locations are not continuously hazardous. They will become hazardous only when a failure of a piping or a ventilation system occurs. The possibility of simultaneous failure of piping and electrical equipment is extremely remote. Where simultaneous failure is possible, it may be necessary to use Class I, Division 1 electrical equipment.

The specific equipment requirements are subject to change during each *National Electrical Code* cycle. The *NEC* is issued every three years. Although there are some changes in Article 501 of the current edition, most of the basic concepts of hazardous location equipment have remained the same over many years. The latest edition of the *NEC* should be consulted for complete requirements. The following paragraphs are a brief summary of the requirements for many pieces of common equipment. See Chapter 4 for more information on the various protection methods discussed.

Switches: Switches are typically arcing devices. Therefore, in Class I, Division 1 locations, they must be installed in explosionproof enclosures or purged enclosures, or they must be intrinsically safe. In Class I,

Division 2 locations, it may be necessary to use a switch rated for Class I, Division 1 locations. This is because in normal operation, the switch may become an ignition source after the area has become hazardous. Division 2 equipment is not permitted to represent an ignition source under normal operating conditions.

Transformers: In a Class I, Division 1 location, transformers containing liquids that will burn must be installed outside of the area in a separate vault. The vault is arranged so that the possibility of communication of gas from the Division 1 location with the transformer vault is minimized. The passage of gas into the vault must be minimized by locating the vault door so that there is no communication between the Division 1 location and the transformer vault. Accumulations of gas or vapors must be minimized by a ventilation system capable of providing continuous removal of flammable gases or vapors. Any ventilation openings and duct openings in the wall of the vault should be in an exterior wall that does not allow communication of gas from the hazardous location. The vent or duct openings must be of sufficient size to relieve any explosion pressures.

Transformers that do not contain liquids that will burn may be installed in the same manner as transformers containing liquids that will burn, or they may be approved for Class I locations. Approved constructions would include explosionproof or purged construction. It is not possible to build power or distribution transformers of intrinsically safe construction due to the high inductances necessary on the primary and secondary of the transformer. For further information, see Section 4-2.1.2.

In Class I, Division 2 locations, a dry-type transformer, a less-flammable liquid-insulated transformer, or a nonflammable fluid-insulated transformer may be installed. However, tap changers, switchgear, and other arcing or sparking equipment must be installed in suitable enclosures. This would imply a need for explosionproof, oil-filled, or purged enclosures for this equipment.

Solenoids: In Class I, Division 1 locations, solenoids must be approved for the location. The most typical approved constructions are explosionproof and intrinsically safe ones.

Explosionproof solenoids are available in a variety of sizes. Intrinsically safe solenoids are limited to smaller sizes. The limitation is due to the larger inductance and higher current that the larger size solenoids need in order to operate. Under fault conditions, the energy stored by the inductance may add to the fault current supplied by the power supply. The combined energy of the power supply plus the inductance may be sufficient to ignite the atmosphere.

In a Class I, Division 2 location, solenoids without arcing or sliding contacts may be installed in general-purpose enclosures. This is because these solenoids do not have contacts capable of igniting a hazardous atmosphere under normal conditions. Where arcing or sparking contacts are present, a general-purpose enclosure may be used if the arcing or sparking contacts are immersed in oil or in an enclosure that is hermetically sealed against the entrance of gases or vapors. A general-purpose enclosure is also permitted if the circuit does not release sufficient energy to ignite the specific atmospheric mixture under normal operating conditions.

Resistors, Reactors, and Heaters: In a Class I, Division 1 location, resistors, reactors, and heaters must be in explosionproof or purged and pressurized enclosures. In a Division 2 location, a general-purpose enclosure may be used if make-and-break or sliding contacts are in explosionproof, oil-filled, or hermetically sealed enclosures, and if the maximum operating temperature of any exposed surface does not exceed 80 percent of the ignition temperature in degrees Celsius of the gas or vapor involved. A general-purpose enclosure is also permitted if the device has been tested and found to be incapable of igniting the gas or vapor.

Motors and Generators: In a Class I, Division 1 location, motors and generators are usually of an explosionproof design. This is the most practical and economic construction available. There are a number of manufacturers producing motors for Class I locations; however, they are generally limited to Group D locations, with a few manufactured for Group C locations. Explosionproof motors and generators are not available for Groups A or B locations. For these installations, it is usually necessary to use a motor that is purged with clean air or some other inert gas. Motors of an intrinsically safe design are only available in the smallest fractional horsepower sizes. Often, the only solution is to locate the motor outside of the Division 1 location.

In a Class I, Division 2 location, the requirements are greatly simplified. If the motor has arcing or switching contacts, integral resistance devices, or centrifugal-type switches, then the motor must be approved for Class I, Division 1 locations, the arcing parts must be suitably protected by explosionproof enclosures, or other protective measures must be taken. Where there are no switching contacts, brushes, or other arcing or sparking mechanisms, an ordinary open or nonexplosionproof enclosed motor, such as a squirrel-cage induction motor, may be used. It is, however, necessary to consider the temperature of internal and external surfaces. These surfaces could be an ignition source if the Division 2 equipment is exposed to a flammable atmosphere.

Portable and Temporary Equipment: The use of portable equipment in hazardous locations is very limited. Most of the portable equipment is used for maintenance purposes, such as portable lighting equipment and portable hand tools. Since maintenance often requires opening housings, a hotwork permit system must be used to declassify the hazardous location to a nonhazardous location. Therefore, some portable hand tools may not need to be suitable for hazardous locations. In some facilities, it is necessary to work on equipment while the area is hazardous. Under these conditions, the equipment under repair should be deenergized, but the maintenance equipment will require hazardous location approval.

Portable lighting equipment used in Class I hazardous locations must be approved for Class I, Division 1 locations. This is because it is difficult to control the area in which it will be used. Portable lights are likely to be taken from the Division 2 to the Division 1 location. In addition, they are often used in maintenance activities that are likely to release ignitible amounts of gases or vapors.

Portable utilization equipment must comply with the same requirements as similar equipment of a fixed nature. The principle difference between portable equipment and fixed equipment is the permissible wiring method. Because of the need for flexibility, a flexible cordset is necessary with most portable equipment. The *NEC* permits the use of flexible cords that are of a type approved for extra-heavy usage.

3-1.1.4 Grounding

There are special grounding requirements for each of the classes of hazardous locations. For Class I locations, requirements can be found in *NEC* Section 501-16. These requirements encompass the basic grounding requirements of Article 250 as well as additional requirements intended to minimize the hazard in the hazardous location. See Section 2-7.

Bonding of piping in process equipment can be critical. Flow of fluid through pipes, hoses, and so forth can generate static electricity. Static charges are equalized when all piping systems and other components are bonded together. If piping systems and other components are not bonded together, significant electrical charges can develop. Rapid dissipation of charge in the fluid can ignite the fluid and cause an explosion. Other static sources may also be present in a room housing this process equipment, which may also represent an ignition hazard. Further information on static electricity can be found in Section 5-1.

Lightning Protection: Lightning may also be a cause of explosions in a hazardous location. Air terminals may be used above an outdoor process area or on the roof of a building to provide a zone of protection against direct lightning strikes. However, large tanks and process towers that are properly grounded do not require lightning rods or air terminals. (See Figure 3-4.) Lightning does not need to strike equipment directly to cause damage. A nearby lightning strike can induce surges in power and communications circuits. Surge arresters may be required within the hazardous location to protect process equipment. Where instrumenta-

Figure 3-4. *A bonding jumper used to ground a large outdoor storage tank to a concrete-encased grid-type electrode.*

tion is connected to field-mounted sensors, it may be necessary to protect the power-supply conductors as well as the signaling line circuits to the field-installed wiring.

Where surge arresters are installed in a Class I, Division 1 location, they must be in suitable (usually explosionproof) enclosures. Where surge arresters are installed in a Division 2 location, no special enclosure is required for the location unless other needs, such as moisture or weather, dictate a need for moistureproof or weatherproof enclosures.

See Section 5-2 for a complete discussion on lightning protection.

3-1.2 Classification of Specific Processes/Occupancies

The following paragraphs discuss occupancies that are typical examples of Class I hazardous locations. Each example is based on the requirements of the *National Electrical Code* and other NFPA standards or recommended practices. Standards and recommended practices are available from other organizations, such as the American Petroleum Institute (API) and the American Gas Association (AGA). The API standards are recommended practices for the classification of petroleum processing plants. The standards of the AGA relate to classification of areas housing equipment for the processing of natural gas. See Section 1-4 for additional information.

3-1.2.1 Automotive Fuel Dispensing Stations

The most frequently encountered hazardous locations are automotive service stations. Until the mid-1970s, most of these service stations were full-service stations where attendants dispensed fuel and performed a number of minor servicing operations. Most of these facilities also included a repair garage for services ranging from minor routine tire changes to major engine overhaul.

During the mid-1970s, many existing stations were converted for self-service usage. Dispensing islands were also added to convenience stores as a sideline business. Public access to, and participation in, the dispensing operation has caused concern regarding fire safety. However, the change from attendant-operated fuel dispensing to self-service has not resulted in a significant increase in the number of fire or explosion incidents.

Gasoline is the most widely used flammable liquid. Since it has a flash point of approximately -40°F (-40°C), it is readily ignitible in ordinary air. In automotive gasoline dispensing stations, the principal concern is that gasoline will be spilled during dispensing or will leak from the tank of the vehicle. The electrical equipment requirements are intended to ensure that the electrical system is not an ignition source. Automotive electrical systems are not required to meet requirements for hazardous locations. In addition, catalytic converters will often have surface temperatures above the autoignition temperature of gasoline. Many have questioned the need for electrical area classification at dispensing islands when those other ignition sources exist. Automotive electrical systems may be eliminated as an ignition source through enforcement of the requirement that engines be shut off during fueling. The catalytic converter is not generally considered an ignition source because no vapor would be in contact with it long enough to be heated to its autoignition temperature. The electrical system of the servicing island must be designed for hazardous locations because vapor may accumulate in some electrical system components. Other components may arc or spark while exposed to flammable vapors. (See Figure 3-5.)

Fuel Dispensing Areas: The area around the dispensing unit is a hazardous location. Therefore, dispensing devices at an automotive service station are located so that all parts of the vehicle being served will be on the premises of the service station. This usually means that a vehicle to be fueled will have to be driven into a parking lot or driveway in order to be serviced. The installation of dispensing units on the sidewalk in front of

Figure 3-5. *Although the vending machine in this photo may be a listed cord-and-plug connected appliance, vending machines of this type are not usually designed or listed for use in hazardous locations.*

the station is prohibited. It is essential that the hazardous location not extend into an area beyond the service station owner's control. Not only does this control ignition sources, but it also decreases the possibility of accidents to the dispenser or to the vehicle being fueled.

The area classification used for the dispensing island depends upon the type of dispenser used. Where an overhead dispenser is used, the dispenser enclosure and all electrical equipment integral with the dispensing hose or nozzle are considered to be located in a Class I, Division 1 location. All space extending 18 in. (457 mm) horizontally in all directions beyond the Division 1 location and extending to grade below the Division 1 location is considered to be a Class I, Division 2 location. A Class I, Division 2 area extends 20 ft (6.1 m) horizontally and 18 in. (457 mm) above grade from the area in which the dispenser is installed. (See Figure 3-6 and Table 3-1.)

Most dispensing units are of the standup type. Some of the area classification within the dispenser is classified as a Class I, Division 1 location. Since dispensing units are required to be listed, the classification of the area within the dispenser is governed by ANSI/UL 87[5].

Gasoline vapors are heavier than air. Therefore, they tend to accumulate in pits and other below-grade locations. For this reason, areas below grade within either the Division 1 or Division 2 location are considered

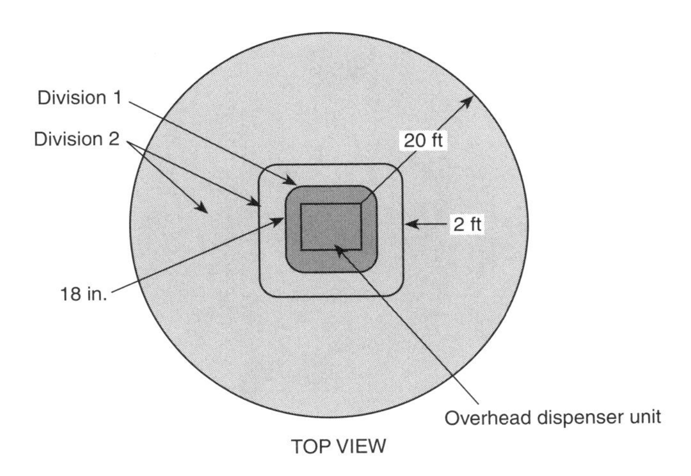

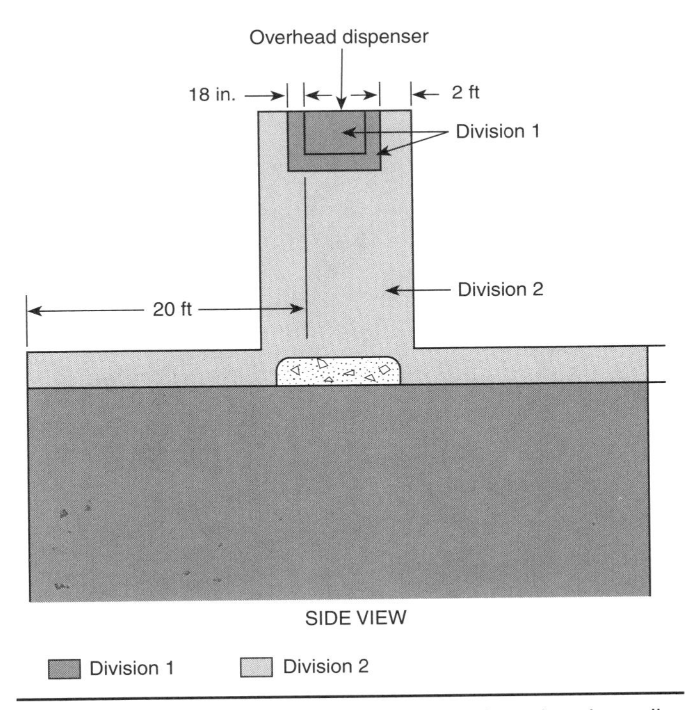

Figure 3-6. *Extent of Class I location around overhead gasoline dispensing units. (Source: National Electrical Code Handbook.[7])*

Table 3-1 Electrical Equipment Classified Areas — Service Stations

Location	*NEC* Class 1, Group D Division	Extent of Classified Area
Underground tank fill opening	1	Any pit, box, or space below grade level, any part of which is within the Division 1 or 2 classified area
	2	Up to 18 in. above grade level within a horizontal radius of 10 ft from a loose fill connection and within a horizontal radius of 5 ft from a tight fill connection
Vent — discharging upward	1	Within 3 ft of open end of vent, extending in all directions
	2	Area between 3 ft and 5 ft of open end of vent, extending in all directions
Dispensing units (except overhead type)*		
Pits	1	Any pit, box, or space below grade level, any part of which is within the Division 1 or 2 classified area
Dispenser	1	Classification inside the dispenser enclosure is covered in power-operated devices for petroleum products, ANSI/UL 87
	2	Within 18 in. horizontally in all directions extending to grade from the Division 1 area within the dispenser enclosure
Outdoor	2	Up to 18 in. above grade level within 20 ft horizontally of any edge of enclosure
Indoor		
With mechanical ventilation	2	Up to 18 in. above grade or floor level within 20 ft horizontally of any edge of enclosure
With gravity ventilation	2	Up to 18 in. above grade or floor level within 25 ft horizontally of any edge of enclosure
Dispensing units, overhead type*	1	Within the dispenser enclosure and all electrical equipment integral with the dispensing hose or nozzle
	2	An area extending 18 in. horizontally in all directions beyond the Division 1 area and extending to grade below this classified area
	2	Up to 18 in. above grade level within 20 ft horizontally measured from a point vertically below the edge of any dispenser enclosure
Remote pump — Outdoor	1	Any pit, box, or space below grade level if any part is within a horizontal distance of 10 ft from any edge of pump
	2	Within 3 ft of any edge of pump, extending in all directions. Also up to 18 in. above grade level within 10 ft horizontally from any edge of pump
Remote pump — Indoor	1	Entire area within any pit
	2	Within 5 ft of any edge of pump, extending in all directions. Also up to 3 ft above floor or grade level within 25 ft horizontally from any edge of pump

Table 3-1 ***(continued)***

Location	*NEC* Class 1, Group D Division	Extent of Classified Area
Lubrication of service room — with dispensing	1	Any pit within any unventilated area
	2	Any pit with ventilation
	2	Area up to 18 in. above floor or grade level and 3 ft horizontally from a lubrication pit
Dispenser for Class 1 liquids	2	Within 3 ft of any fill or dispensing point, extending in all directions
Lubrication or service room — without dispensing	2	Entire area within any pit used for lubrication or similar services where Class 1 liquids may be released
	2	Area up to 18 in. above any such pit, and extending a distance of 3 ft horizontally from any edge of the pit
	Nonclassified	Any pit, below-grade work area, or subfloor work area that is provided with exhaust ventilation at a rate of not less then 1 ft^3/minute · ft^2 (0.3 m^3/minute · m^2) of floor area at all times that the building is occupied or when vehicles are parked in or over this area and where exhaust air is taken from a point within 12 in. (0.3 m) of floor of the pit, below-grade work area, or subfloor work area
Special enclosure inside building for tanks**	1	Entire enclosure
Sales, storage, and rest rooms	Nonclassified	If there is any opening to these rooms within the extent of a Division 1 area, the entire room must be classified as Division 1
Vapor processing systems pits	1	Any pit, box, or space below grade level, any part of which is within a Division 1 or 2 classified area or that houses any equipment used to transport or process vapors
Vapor processing equipment located within protective enclosures	2	Within any protective enclosure housing vapor processing equipment
Vapor processing equipment not within protective enclosures (excluding piping and combustion devices)	2	The space within 18 in. in all directions of equipment containing flammable vapor or liquid extending to grade level. Up to 18 in. above grade level within 10 ft horizontally of the vapor processing equipment
Equipment enclosures	1	Any area within the enclosure where vapor or liquid is present under normal operating conditions
Vacuum-assist blowers	2	The space within 18 in. in all directions extending to grade level. Up to 18 in. above grade level within 10 ft horizontally

Source: NFPA 30A, Automotive and Marine Service Station Code[6].

**Ceiling-mounted hose reel.*

***See Section 2-2 in NFPA 30A for information on special enclosures for tanks.*

For SI units: 1 in. = 2.5 cm; 1 ft = 0.30 m.

NOTE: The area classifications listed in Table 3-1 are based on the premise that the installation meets all of the applicable requirements of NFPA 30A in all respects. If this is not the case, the authority having jurisdiction has the authority to determine the extent of the classified area.

to be Class I, Division 1 locations. This includes any box or equipment that is installed flush with grade or below grade. The area 20 ft (6.5 m) horizontally and 18 in. (457 mm) above grade from the exterior of the dispenser is considered to be a Class I, Division 2 location. However, any pit or depression within this 20-ft (6.1-m) area is a Class I, Division 1 location, because vapors will accumulate and will not be readily dissipated.

Remote pumps are often used to pump gasoline from the storage tank to the dispensing unit. These remote pumps are usually installed outdoors, below grade in a pit. One pump may supply several dispensers with the same grade of gasoline. The below-grade space for a distance of 10 ft (3.05 m) from the edge of the pump is considered to be a Division 1 location, because vapors cannot escape the pit. The area within 3 ft (914 mm) of the edge of any pump, extending in all directions, is considered to be a Division 2 location. Up to 18 in. (457 mm) above grade level within 10 ft (3.05 m) horizontally from the edge of the pump is also considered to be a Division 2 location. To ensure that the hazardous location does not extend to an area beyond the control of the service station owner, the pump must be installed at least 10 ft (3.05 m) from adjoining property that can be built upon. NFPA 30A requires the pump to be installed at least 5 ft (1.52 m) from any building opening. NFPA 30A also requires dispensers and pumps to be listed.

Indoor locations for dispensing units are permitted but are less desirable, because they require a Division 2 classification within 5 ft (1.52 m) in any direction from the pump, and within 25 ft (7.62 m) horizontally for a height of 3 ft (914 mm) above the floor. Typically they employ explosionproof or intrinsically safe equipment. They may also utilize a combination of protection methods.

Where modification of the dispensing unit has become necessary due to circumstances, such as a pricing increase beyond the capability of the fuel totalizer, it may be necessary to modify the unit. Modification of listed equipment often voids the product listing. Some laboratories, such as Underwriters Laboratories Inc., may permit modification where the modification to be made has been "listed by report." This is a special form of listing employed that covers products or constructions for which there are no generally recognized installation requirements. Information concerning proper installation should be contained in a report identified by the reference and date shown by the listing, copies of which can be obtained from the listing organization.

The authority having jurisdiction may grant permission to modify a fuel dispensing device. However, the listing agency will usually consider such a modification to void the listing.

Vapor Recovery Systems: In many areas of the United States, systems to recover gasoline vapors are required at automotive service stations.

At present, laws regarding the recovery of gasoline vapors are being enforced in two phases. Phase I requires recovery of the vapor when making a delivery to an underground tank. Phase II requires the recovery of gasoline vapors when an automobile is being fueled, and it involves the capture of vapors being expelled from the automobile fuel tank as the liquid fills the tank and displaces the vapors.

At present, there are two types of Phase II systems being used for vapor capture at automobile service stations: the vapor recovery system and the vapor processing system.

The vapor recovery system captures, and retains without processing, flammable liquid vapors displaced during the filling of tanks or containers or during the fueling of the vehicles. This can be accomplished through systems such as balance pressure displacement systems, or so-called hybrid systems. Balance pressure displacement systems return the gasoline vapors through vapor-return hoses and piping back to underground tanks. The hybrid systems return some of the gasoline through vapor-return lines at the pumps before the gasoline reaches the meters. This action draws a vacuum on the gasoline tank being filled, causing vapors to return through the vapor-return line.

Vapor processing captures and processes the flammable liquid vapors displaced during the filling of tanks or containers or when filling vehicles. The vapors are captured by the use of mechanical and/or chemical means through systems that use blower-assist, and then are processed through refrigeration, adsorption, or combustion. Generally, those systems using refrigeration, and in some cases those using adsorption, are found in large bulk plants.

The area around either type of vapor capture system is usually considered to be classified electrically. Figure 3-7 illustrates the two common vapor capture systems and their related equipment. The vicinity of the vapor recovery equipment is considered to be in a Class I, Division 2 location. This classification extends 18 in. (457 mm) in all directions from the equipment containing the vapor or liquid, extending to grade level. It also extends 10 ft (3.05 m) horizontally from the equipment up to a level of 18 in. (457 mm) above grade.

Vapor recovery systems are usually constructed of explosionproof and intrinsically safe equipment. However, some purging systems are also used.

Underground Storage Tanks: Tanks at service stations are generally required to be installed underground. These buried tanks normally have a capacity in excess of 5000 U.S. gallons (18,925 L), with a vent pipe extending vertically from the tank.

It is important that the vent pipe not have U-bends or weatherhoods designed to keep out rain and foreign matter. Experience has shown that the risk of serious contamination of the tank contents that can accompany

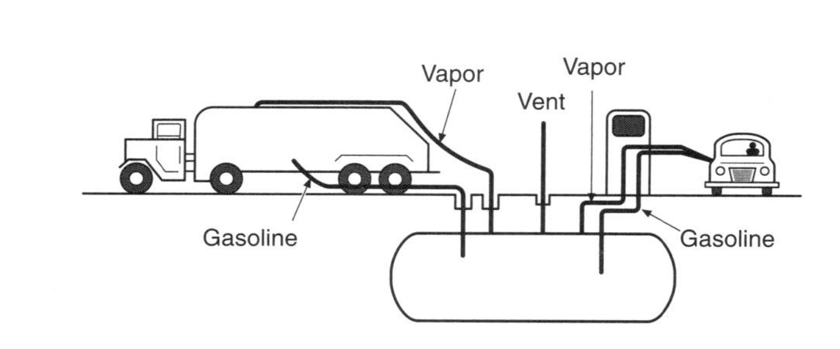

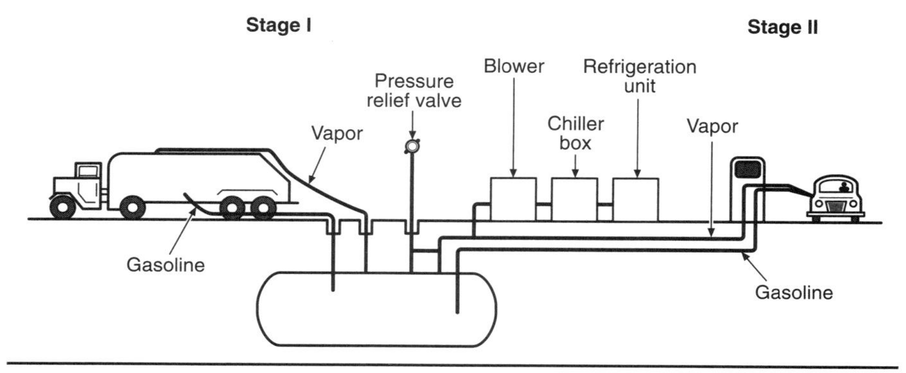

Figure 3-7. *(Top) Vapor recovery system. (Bottom) Vapor processing system. (Source: NFPA 30A.)*

upward discharge of vapor from an open vent is small. However, there have been several serious incidents of fire when a U-bend or weatherhood caused the vapors to be directed downward. Such an installation should be corrected to comply with NFPA 30A. With an upward-discharging vent, a Division 1 location exists within 3 ft (914 mm) of the open end of the vent, extending in all directions. The area between 3 ft (914 mm) and 5 ft (1.52 m) of the open end is normally considered to be a Division 2 location.

Below-grade spaces within a 10-ft (3.05-m) radius of an underground tank are considered to be Division 1 locations. The fill opening of the tank is normally considered to be a Division 1 location. At grade level, the Division 1 classification does not extend beyond the fill opening of the tank. A Division 2 location extends above grade. The extent of the Division 2 location is based on the tightness of the fill connection. Where the connection is loose, the Division 2 location extends 18 in. (457 mm) above grade for a 10-ft (3.05-m) radius around the fill opening. Where the connection is tight, the Division 2 location extends 18 in. (457 mm) above grade for a 5-ft (1.52-m) radius around the fill opening. The reason for the more stringent classification below grade level is that gasoline vapors will tend to accumulate in underground locations without forced-air ventilation. Such locations, therefore, must be considered Division 1 locations.

Vehicle Repair Areas: Many gasoline dispensing stations include a vehicle repair area. However, the percentage of stations with repair facilities is significantly lower than a few years ago. Many new facilities provide fuel service only or repair services only. Except in unusual circumstances, the classification of repair areas in gasoline dispensing stations is the same as the classification of repair areas in commercial garages. See NFPA 88B, *Standard for Repair Garages*[8], for classification information.

3-1.2.2 Marine Fuel Dispensing Stations

Marine fuel dispensing stations are similar in many ways to automotive fuel dispensing stations. The typical fuels used are gasoline and kerosene (diesel fuel). There are, however, some unique problems associated with marine fuel dispensing. The fuel dispensers are usually located on open piers, on wharves, on floating docks, or on shore. The pump may be integral with the dispensing unit. However, some are not. Tanks that are not integral with the dispensing device are normally required to be on shore or on a pier of solid-type fill. Where a shore location would require an excessively long supply line to the dispenser, the authority having jurisdiction may authorize installation of tanks on a pier. Where installed on a shore location, a high water table or a rocky shoreline may make it impractical to bury a storage tank. In some installations, the tank may be higher in elevation than the dispenser, resulting in a pressure head at the dispenser. A break in the fill line has the potential to drain the tank, unless a solenoid shutoff valve is installed in the fill line. Aboveground tanks are less desirable than buried tanks, because they may be exposed to external fires and are more prone to vehicle accidents.

Corrosion is a significant problem at marine fuel dispensing stations. This is particularly troublesome at saltwater marine fuel terminals. Salt can attack enclosures, conduit, wiring, terminal strips, piping, fittings, and so on. Corrosion can cause discontinuity in the equipment grounding path of conduit systems. It can also lead to stray voltages in equipment, which may be hazardous to personnel. Equipment in such hazardous locations must be corrosion resistant in order to ensure that the protection method will be effective. No electrical equipment should be installed in any area where it is likely to be submerged, unless it is specifically intended for the purpose.

Dock space at marinas is generally very costly, and this has caused developers to over-subdivide these areas. Therefore, a number of activities may be located very close to the fuel dispensing area. It is not uncommon to find ordinary receptacles, flexible cords, and unsuitable electrical equipment located within the hazardous location. It is important to ensure that no electrical equipment unrelated to fuel dispensing is located in the hazardous location.

The principal concern in marine fuel service stations is a spill of gasoline. Many of the hazards are similar to those in an automotive fuel dispensing station. Table 3-1 should also be used to identify classifications of the hazardous areas in a marine fuel dispensing station.

Marine fuel dispensing presents some additional problems beyond those in an ordinary gasoline station. Gasoline spills on a boat are particularly hazardous, because the vapors will tend to accumulate at the bottom of the boat. Blowers are provided in some boats to purge the bilges of vapors. When blowers are used according to instructions, and safe dispensing practices are followed, marine fueling can be conducted safely. These safe fueling practices are beyond the scope of this handbook.

NFPA 30A provides the requirements for fuel servicing stations. Specific requirements for boats can be found in NFPA 302, *Fire Protection Standard for Pleasure and Commercial Motor Craft*[9].

3-1.2.3 Commercial Garages — Repair and Storage

Repair garages usually have large quantities of combustible oils. However, they are usually below their flash point at room temperature. The principal concern is a leak of gasoline from a vehicle. As indicated in Section 3-1.2.1, gasoline is readily ignitible in ambient air. Some repair stations introduce an additional hazard by using gasoline as a solvent to clean automotive parts. This is a dangerous practice that may require more stringent classification than described in this section. The section on dipping and coating may be used as a guideline for classification of those areas. However, the practice should be discouraged.

Article 511 of the *National Electrical Code* applies to some garages used for vehicle service and repairs. The requirements apply to vehicle storage areas using electrical equipment, open flames, or welding as part of the repair operation. They also apply to an area where volatile flammable liquids may be stored. An area in which the repair work consists of routine exchange of parts, without the need for electrical equipment, is not normally considered hazardous. Parking garages and automobile dealer showrooms are also not likely to be considered hazardous, even though these showrooms will normally have some form of electrical equipment. The risk of explosion in one of these areas is minimal, since maintenance work on the fuel tank is not being performed. (See Table 3-2.)

Table 3-2 Class I Locations — Commercial Garages Repair and Storage

Location	Class 1, Group D Division	Extent of Class I Location
1. Fuel transfer or dispensing area	See Table 3-1	
2. Parking garages	Nonhazardous	Applies to areas in which no fuel transfer or maintenance is performed, except exchange of parts and routine maintenance requiring no electric equipment, welding, open flame, or volatile flammable liquid
3. Up to 18 in. above the floor	Division 2	Vehicle repair areas not covered by Item #2 above
4. Pit or depression below floor in a Class I location	Division 1	Extends from the bottom of the pit to floor level
5. Pit or depression below floor in a Class I location when the ventilation rate in the pit is six air changes per hour	Division 2	Extends from the bottom of the pit to floor level
6. Stock rooms, switchboard rooms, offices, rest rooms, and so on	Nonhazardous	Applies to areas that are ventilated at the rate of at least four air changes per hour or are effectively cut off by walls or partitions
7. Areas adjacent to Class I locations	Nonhazardous (by special permission)	Applies to areas that, in the opinion of the authority having jurisdiction, have sufficient ventilation, pressure differential, or space separation
8. Automobile dealer showrooms	Nonhazardous	Applies to dealer showrooms where no vehicle maintenance is performed (except as covered in Item #2 above)

Some parking garages do require classification, particularly if fuel is dispensed within the parking garage. Specific requirements for the fuel dispensing can be found in NFPA 30A. Further information on parking garages can be found in NFPA 88A, *Standard for Parking Structures*[10]. Although parking garages used strictly for vehicle storage are not considered as classified under Article 511 of the *National Electrical Code*, adequate ventilation is required to carry off exhaust fumes from the engine.

Servicing areas in a commercial garage require classification. The area 18 in. (457 mm) above the floor is classified as Class I, Division 2, Group D. The classified area around dispensing drums for flammable liquids or fuel dispensing pump locations must be classified based on these hazards. (See Section 3-1.2.1 for fuel dispensing areas; see Section 3-1.2.11 for dispensing of flammable liquids from drums.) Pits or depressions are usually classified as Class I, Division 1, Group D locations. Hydraulic lifts have replaced servicing pits in general repair garages. However, pits are becoming common in lubritoriums that provide express oil changes and lubrication. In these locations fuel is not dispensed, and it is not expected that fuel systems will be serviced or repaired. Large concentrations of fuel vapors do not accumulate, allowing pits in lubritoriums to be classified as Class I, Division 2, Group D locations. Many lubritoriums provide ventilation in the pit. Pits or below-grade work areas may remain nonclassified where ventilation is provided at a rate of not less than 1 ft^3/ minute · ft^2 (0.3 m^3/minute · m^2) of floor area when the building is occupied. The exhaust air should be taken from a point within 12 in. (0.3 m) off the floor of the pit or below-grade work area.

In any garage there will usually be electrical equipment above the hazardous location. Article 511 of the *National Electrical Code* contains some rules for equipment and wiring in spaces above the Class I location. The basic concept for these applications is that equipment installed above the hazardous location must be designed such that any arcs, sparks, or particles of hot metal are contained by the equipment enclosure. Fixed lighting must be located 12 ft (3.66 mm) or more above the floor level, or of the totally enclosed type, or constructed so as to prevent the escape of sparks or hot metal particles.

Wiring above a Class I location must be in metal raceways, rigid nonmetallic conduit, or electrical nonmetallic tubing, or it must be Type MI, TC, SNM, or Type MC cable. Cellular metal floor raceways or cellular concrete floor raceways are permitted to supply ceiling outlets or extensions to the area below the floor, but such raceways are not permitted to have any connections leading into or through any Class I location above the floor.

Auto body shops usually have paint-spraying facilities. The paint spraying may be conducted in the open or performed in a booth. Paint spraying is a hazard that is not considered in the requirements of Article 511 of the *NEC*. The requirements can be found in Article 516. For further information, see Section 3-1.2.6.

3-1.2.4 Aircraft Fuel Dispensing

There are three basic types of aviation fuels: aviation gasoline, kerosene, and blends of gasoline and kerosene.

Aviation gasoline (AVGAS) is limited primarily to smaller, general aviation aircraft, including airplanes and helicopters. It is similar in many respects to automotive gasoline. In fact, many aircraft are now using automotive-grade gasolines. Of all of the aviation fuels, AVGAS is the most flammable.

There are several grades of kerosene in common use in turbine-powered aircraft. JET A and JET A-1 are in common general aviation and airline usage. Aircraft using such fuels include turbo-props, helicopters, and jets. JET A is the least flammable of the common aviation fuels.

JP-5 and JP-6 are military-grade kerosene fuels that have characteristics similar to those of JET A.

JET B and JP-4 are blends of gasoline and kerosene. JP-4 is the most frequently used fuel in turbine-powered military aircraft. When an aircraft type may be used for military or civilian purposes, the military users may use JP-4 in lieu of JET A depending on availability. JET B is the civilian equivalent to JP-4.

AVGAS, JET B, and JP-4 are the most hazardous types of aviation fuels, because they have the lowest flash points. The flash point of standard grades of aviation gasoline has been established at approximately -50°F (-46°C) at sea level by the tag closed-cup method. The flash point of JET B turbine fuel is not regulated by specification, but samples have been tested by the closed-cup method and have been found to be in the range of -10°F to 30°F (-23°C to -1°C) at sea level. Most of the JET A (kerosene-grade) turbine fuels have flash points in the range of 95°F to 145°F (35°C to 63°C) (closed-cup) at sea level.

JET A (kerosene grade) turbine fuels offer a safety advantage over other types of aviation fuels. It is not as easily ignitible at normal sea level ambient as other types of aviation fuels, due to a higher flash point [generally above 100°F (38°C)]. In some areas of the world, JET A is not readily available, so JET B or JP-4 must be used. JET B is more economical to produce. Some nations have proposed a decrease in the flash point to increase the yield of fuel per barrel. The minimum proposed flash point of this reduced flash point kerosene (RFK) is in the range of 80°F (27°C). However, most of the samples tested had flash points of 100°F (38°C). If the fuel shortages of the 1970's return, increased use of RFK and JET B can be expected.

Aviation gasoline and JET B turbine fuel at normal temperatures and pressures will give off vapors capable of forming ignitible mixtures with the air near the surface of the liquid or within the vessel in which the liquid is stored. Kerosene grades of fuels (JET A) will not produce ignitible vapors at normal temperatures and pressures, but where a JET A fuel may be heated above its flash point, the vapors may be capable of forming ignitible mixtures. This condition may develop where ambient temperatures are in the 100°F (38°C) range for extended periods, such as in tropical regions. The kerosene-based fuels may be explosive in air at normal temperatures if released as a mist. This will normally happen only if the fluid is under pressure. Such a condition could occur during fueling operations.

Most airports dispense several grades of aviation gasoline and one or more of the grades of turbine fuel. The most commonly encountered turbine fuels at civilian airports are JET A and JET A-1. Because of the likelihood that any grade of aviation fuel may be encountered, it is necessary to classify fuel servicing locations and aircraft hangars for the more flammable aviation gasolines.

The type of dispensing system used will usually depend on the size and amount of air traffic using the airport. The smallest low-traffic airports may use drums and drum pumps to fuel aircraft. Some of the slightly larger airports dispense from fixed-fuel dispensers. Mid- to large-size airports usually use tank truck dispensing. The use of aboveground fuel dispensers is generally limited to gasoline dispensing for reciprocating engine general aviation aircraft. Below-grade fuel dispensers are occasionally used for commercial or military aviation purposes. This method removes an obstruction that could damage aircraft wings. The specific requirements for aircraft fuel servicing can be found in NFPA 407, *Standard for Aircraft Fuel Servicing*[11].

Where an aircraft is fueled from an outside aboveground fuel dispenser, some of the space inside the fuel dispenser and all below-grade spaces are considered to be a Class I, Division 1 location. A Class I, Division 2 location extends from the edge of the dispenser for a radius of 20 ft (6.1 m) from the dispenser. The classification scheme for areas in which gasoline is the fuel being dispensed can be found in Article 514 of the *National Electrical Code*. A summary of these requirements can be found in Table 3-1.

When an aircraft is fueled from a below-grade dispenser, the fuel dispenser is considered to be in a Division 1 location. This classification is necessary because vapors tend to accumulate in the pit in which the dispenser is located. A Division 2 location is considered to extend 20 ft (6.1 m) in all directions from the fuel dispenser. This method of dispensing is generally limited to larger turbine-powered aircraft.

Static electricity is a significant concern while fueling aircraft. The degree to which static charges are acquired depends on residual impurities, dissolved water, the linear velocity through piping systems, and the types of filters and water separators used. Static development can also be significant due to the large quantities of fuel transferred. The kerosene grades of fuel can generate more static charge than AVGAS.

Static electricity can be managed if static charges are dissipated before they reach dangerous levels. Charges can be equalized through conductive dispensing hoses and through bonding and grounding. (See Figure 3-8.) The aircraft and the dispenser must also be bonded together. This is usually done through a flexible cable with alligator clips attached to a conductive part of the aircraft that is in electrical contact with the aircraft fuel tank. Where fuel is dispensed from permanently installed dispensers, the dispenser should be permanently connected directly to a grounding electrode. Where fuel servicing is from dispensing trucks, a flexible cable should be run from the dispensing vehicle to a grounding electrode. Airports where tank truck dispensing is common usually have readily accessible electrodes for grounding. These electrodes may also be designed to be used as tie-down points for aircraft parking restraint. Further information on static electricity can be found in Chapter 5.

Figure 3-8. *Aircraft fuel dispensing from an underground tank using a mobile dispensing truck. Note the use of the bonding jumpers to the aircraft landing gear, to the grounding electrode at the dispenser, to the coupling at the tank, and to the coupling at the aircraft.*

Fuel must be stored in bulk quantities for dispensing at small- to midsize general aviation airports using fuel dispensers. Fuel is often stored in buried fuel tanks near dispensing units. At facilities using tank truck distribution and manifolded pit distribution, a bulk fuel distribution facility is often located at the airport. It may consist of a few small tanks, or it may be a large tank farm facility. Article 515 of the *National Electrical Code* contains requirements for classification of bulk fuel storage facilities. A summary of the requirements can be found in Table 3-3.

3-1.2.5 Aircraft Hangars

An aircraft hangar is a building or other structure in which aircraft are stored and in which aircraft may undergo servicing, repairs, or alterations. There are three general classifications of hangars, which are denoted by groups.

A Group I hangar is a hangar with at least one of the following conditions:

1. An aircraft access door height over 28 ft (8.5 m)
2. A single fire area in excess of 40,000 ft^2 (3716 m^2)

Table 3-3 Classification of Aircraft Hangars

Location	Class 1, Group D	Extent of Location
Hangar floor	Division 2	Entire area of hangar from floor up to 18 in., including any areas not cut off or ventilated
Pits or depressions	Division 1	Any below-grade space
Vicinity of aircraft	Division 2	5 ft horizontally from any power plants, extending 5 ft vertically above the engine enclosure to the floor
	Division 2	5 ft horizontally from any fuel tanks that extend 5 ft vertically above the upper surface of the wings and to the floor
Stock rooms, electrical equipment rooms, and so on	Nonhazardous	Applies to rooms that are adequately ventilated and suitably cut off

3. Housing aircraft with a tail height over 28 ft (8.5 m)
4. Housing strategically important military aircraft as determined by the U.S. Department of Defense

A Group II aircraft hangar is a hangar in which an aircraft access door has a height of 28 ft (8.5 m) or less and a single fire area that is defined by building construction type and building area.

A Group III aircraft hangar may be a free-standing individual unit for a single aircraft, a row hangar having a common structural wall and roof system and housing multiple aircraft as well as having door openings for each aircraft, or an open-bay hangar capable of housing multiple aircraft and having an aircraft access door height of 28 ft (8.5 m) or less.

Group I hangars are often used as maintenance hangars. Group II and Group III hangars are typically used primarily for parking. However, aircraft maintenance is usually performed in all hangars. The large Group I hangars are used by commercial airlines and military organizations for heavy maintenance of aircraft.

The principal concern in an aircraft hangar is a spill of fuel from a fuel tank or a fuel line. For most general aviation aircraft, the likelihood of a fuel leak is high, because a small amount of fuel is routinely drained from sumps in aircraft as part of a preflight inspection. (See Figure 3-9.) A typical small aircraft will have at least three fuel drains. Fuel is drained from each sump to check for the presence of water in the fuel. When water is detected, the drain is operated until the water is removed and only fuel is present. Unfortunately, many of these preflight inspections of general aviation aircraft are conducted in hangars.

Figure 3-9. *Fuel drains on a general aviation aircraft.*

Large transport aircraft contain large quantities of combustible fuel. Transport aircraft are generally in hangars for maintenance only. The frequency of maintenance will vary. Operations may be performed all at once in a single visit or as part of a progressive overhaul system where a certain portion of a total overhaul is conducted periodically. In addition, periodic inspections are required to detect minor maintenance discrepancies. Since progressive maintenance of large aircraft in airline fleets is continuous, it is obviously impractical and economically unsound to remove all fuel from an aircraft prior to moving it into a hangar. Therefore, the potential always exists for having large quantities of flammable aviation fuel inside a hangar. In addition, some maintenance procedures involve the use of flammable solvents, sometimes-unstable or toxic chemicals, or personnel entry into fuel tanks.

Area classification for a hangar must be based on the worst-case aviation fuels (i.e., AVGAS or JP-4). This is because a hangar could be used to house aircraft using any of the aviation fuels. Major hull repair often requires repainting of the aircraft. Some hangars are designed especially for spray painting. However, some spray painting of general aviation aircraft is done in ordinary hangars. Spray painting hazards are covered by NFPA 33, *Standard for Spray Application Using Flammable or Combustible Materials*[12], and Article 516 of the *NEC*.

Table 3-4 summarizes the classification of areas in a hangar in accordance with Article 513 of the *NEC*. Figure 3-10 illustrates how these requirements may be applied to a typical hangar containing a large transport-category jet.

Table 3-4 Class I Locations for Dipping and Coating Processes

Location	Classification	Extent of Location
Dip tank	Class I, Division 1	5 ft (1.5 m) radially in all directions from vapor source extending to the floor
	Class I, Division 2	5 ft (1.5 m) to 8 ft (2.5 m) in all directions from vapor source. 25 ft (7.5 m) horizontally and 3 ft vertically above the floor
Drainboard	Class I, Division 1	5 ft (1.5 m) radially in all directions from the drainboard extending to the floor
	Class I, Division 2	5 ft (1.5 m) to 8 ft (2.5 m) in all directions from vapor source or wetted surface, extending to the floor. 25 ft (7.5 m) horizontally and for 3 ft (1 m) above the floor
Pits (within 25 ft)	Class I, Division 1	Any pit or space below grade level, any part of which is below grade level. If the pit extends beyond 25 ft, a vapor stop should be provided. Otherwise, the entire pit is classified as Class I, Division 1.
Enclosed processes	Class I, Division 1	Within the process enclosure
	Nonhazardous	Process oven. Area beyond enclosed process equipment.
	Class I, Division 2	In vestibule 3 ft (1 m) from the Class 1, Division 1 location. Within 3 ft (1 m) in all directions from openings.

The entire floor area of the hangar, including any adjacent and communicating areas that are not suitably cut off, is classified as a Class I, Division 2 location extending 18 in. (457 mm) above the floor. The area within 5 ft (1.52 m) horizontally from aircraft power plants or aircraft fuel tanks extending from the floor upward to a level 5 ft (1.52 m) above the upper surfaces of the wings and engine enclosures is also classified as a Class I, Division 2 location. Pits or depressions below the floor are classified as Class I, Division 1 locations up to the floor surface.

In order to properly classify the area, it is necessary to obtain information on aircraft parking patterns, the types of aircraft, and the operations to be performed in the hangar. Hangars used to service airline fleets are usually designed for definite parking patterns. Figure 3-11 illustrates two maintenance areas of an airline hanger.

Solvents are often used in maintenance hangars. Solvent storage and dispensing may require classification beyond that already required for the hangar. For further information, see Section 3-1.2.11.

Military aircraft may pose special hazards requiring special precautions. These hazards include canopy ejectors, armament rockets, pyrotechnics, rocket propellants, and special high-energy fuel. Components of these systems require special storage and handling. If the hazardous

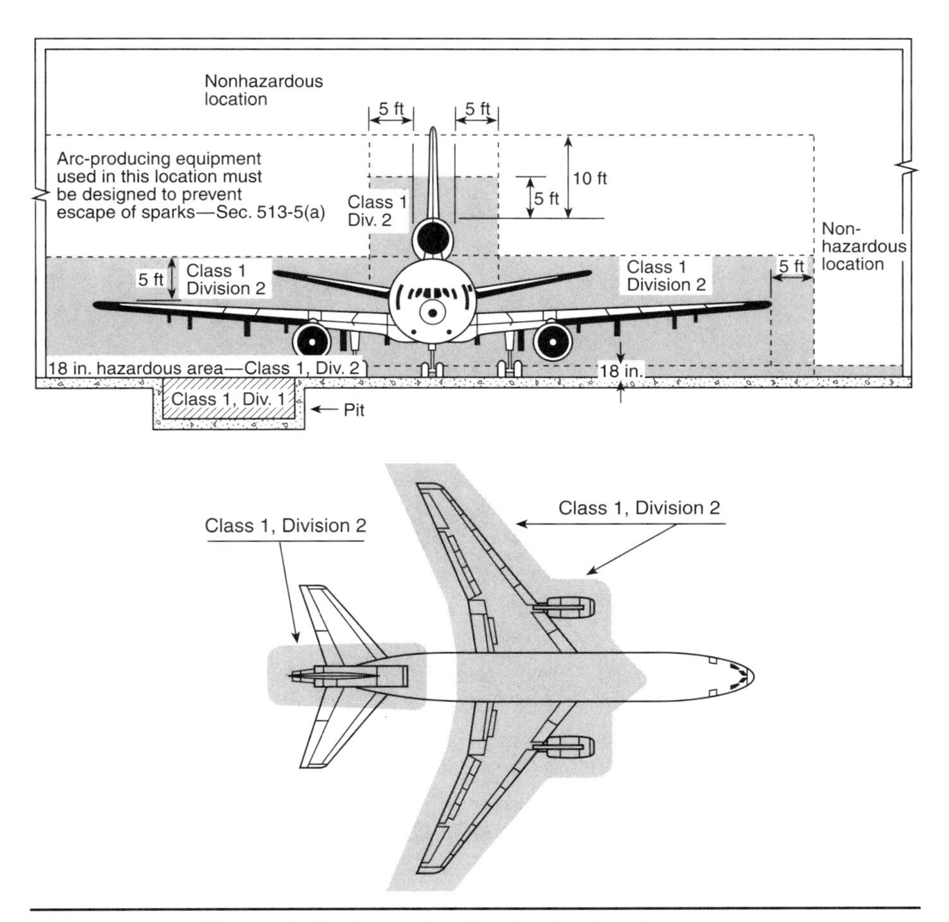

Figure 3-10. *Area classification in aircraft hangars.*

location group of the particular agent involved is different from that required for aviation fuels (Group D), then the area classification for those components of the electrical system likely to be exposed to both groups must be suitable for exposure to both materials.

Maintenance of large aircraft often requires the use of stanchions or rostrums to facilitate access. (See Figure 3-11.) The need for area classification around this work platform is dictated by the process involved and the proximity to aircraft fuel systems or power plants.

Article 513 of the *NEC* provides additional electrical requirements beyond those in Article 501. The requirements address the following three distinct areas:

1. Any room that is suitably cut off from the main hangar area
2. The Class I area near the floor engines

Figure 3-11. *Two typical service areas in a Type 1 hangar used for airline maintenance purposes. The top photo depicts an area in which general maintenance is performed. The wall toward the back of the hangar has a corridor through which the nose of the aircraft protrudes. The bottom photo depicts a maintenance dock used for more elaborate airframe maintenance. The parking patterns necessitated by these construction features simplify area classification.*

3. Fuel tanks and the nonhazardous location above the floor

In nonhazardous areas that are suitably cut off from the main hangar area, the wiring requirements of Chapter 3 of the *NEC* (rather than the more stringent hazardous location requirements of Chapter 5) apply. In the main hangar area, the wiring in the Class I location must comply with the requirements of Article 501. In the nonhazardous location above the hazardous location, wiring must be installed in metallic raceways or be Type MI, TC, SNM, or MC.

Equipment that is installed less than 10 ft (3.05 m) above wings and engine enclosures of aircraft should preferably be of a design that does not include arcing or switching contacts. As an alternate, the equipment should be of the totally enclosed type and so constructed as to prevent the escape of arcs, sparks, or hot metal particles. (See Figure 3-8.)

The *NEC* requires portable lighting equipment to be approved for the location in which it is to be used. This does not mandate the use of lighting equipment suitable for a Class I, Division 2 location if it is not going to be used in the Class I area. However, most portable lighting equipment used in a hangar will most likely at some time be used in the Class I locations.

All other portable utilization equipment that is used in a hangar must be approved for Class I, Division 2 hazardous locations. This is because the equipment is likely to be used in the hazardous location. Any portable equipment used in hangars must have flexible cords that are approved for extra-hard usage. A listing of flexible cord uses can be found in Article 400 of the *National Electrical Code*.

3-1.2.6 Spray Application of Flammable and Combustible Materials

Types of Spray Application Systems: Spray application of coatings is a common practice in industrial facilities, ranging from auto body shops to large assembly lines. The process may involve a solvent-based fluid material, such as paint, or a solvent-free powder coating material. The most common atomizing device for fluid coatings is the air-spray gun. The air-spray gun uses high-pressure jets of air to break up the fluid into a high-pressure mist. Another common fluid-coating gun, the airless atomizer, sprays by hydraulic means without compressed air. Fluid pressures in airless atomizers can range from 300 psi (2068 kPa) to approximately 3000 psi (20,685 kPa).

Air and airless atomizers are also used in electrostatic spray operations. An electrostatic spray gun uses an atomizing head with a high-voltage electrical input that charges the paint droplets. Voltages applied to the gun range from 35,000 V to over 100,000 V. There is less overspray with electrostatic paint spraying, because the charged atomized particles are attracted to the workpiece. This results in less cleanup. It also makes the electrostatic gun more economical to use.

Other types of electrostatic atomizers include disk-type and bell-type atomizers. The disk-type atomizer consists of a vertically mounted spinning disk, which is charged to approximately 100 kV. The coating fluid is poured onto the spinning disk which then spreads the fluid on the workpiece in a 360-degree pattern. The bell-type atomizer is a disk atomizer formed in the shape of a cup or bell. The charged bell rotates at very high speeds. Bell atomizers are often mounted in banks.

Powder coating is a process in which a dry organic powder is suspended in air and charged electrostatically from a high-voltage dc power supply. The powder is directed toward the workpiece and is held in place by the electrostatic charge. The workpiece is then passed into a process oven, which melts the powder into a uniform coating. Powder coatings may be applied with spray guns that mix powder with air. A cable connected to the gun supplies high voltage for the electrostatic charge.

In addition, other methods of applying powder coatings may be employed. For small pieces having dimensions of less than 4 in. (102 mm), an electrostatic fluidized bed may be used. In this process, powder is held in a container having an open top and a porous bottom through which an upward flow of air causes the powder mass to levitate or "fluidize." Charging electrodes near the surface impart an electrical charge to the powder. Electrically grounded workpieces pass over the surface of the bed and collect a coating of powder. The coating is then melted into a uniform coating and then cured in a baking oven.

Although they are not widely used, there are techniques for applying dry powder coatings without the use of electrostatics. These processes typically involve preheating the workpiece to a temperature substantially above the melting point of the powder and then applying the powder, either by dipping the workpiece into a fluidized bed or by spraying air-suspended powder directly onto the hot surface of the workpiece. The powder melts immediately upon contact and flows to form a film, which is subsequently cured in a bake oven.

Spray Area: Most spraying operations are performed in a spray booth. A booth not only limits the extent of overspray, but it also can limit the extent of a hazardous location. There are several types of spray booths: the open-faced booth, the enclosed booth, and the tunnel booth.

The most common type of spray booth is the open-front or open-face type, which is a box-like structure with one side open. A ventilation system is provided to dissipate the flammable vapors. For assembly-line-type operations, a conveyor line opening may be provided for work entry and exit through the booth.

Tunnel-type booths are often used on assembly lines for larger mass production facilities, such as automobile assembly plants. These booths are long booths that often have more than one spray station. A conveyor in the floor transports the workpieces within the booth. Vertical down-draft ventilation is usually used to minimize flammable vapor accumulation.

Enclosed spray booths are used where workpieces can be brought into a booth that has its doors closed. (See Figure 3-12.) Spraying in enclosed booths is usually performed with manual equipment. A ventilation system is installed to reduce the vapor concentration during spray and cleaning operations.

Continuous coaters are a special type of spray booth engineered for a process. The coater consists of a booth containing banks of spray guns. Entrance and exit vestibules are provided with mechanical ventilation, which minimizes the escape of flammable vapors from the booth.

In some operations, it is not practical to perform all spraying within a booth or room. It may be necessary to conduct spraying in a general factory operating area. This is known as open-floor spraying. Open spraying operations are not generally desirable, because they require a large buffer zone to minimize the likelihood of ignition of overspray from any number of ignition sources.

NFPA 33 and Article 516 of the *National Electrical Code* provide requirements for the classification of spray application areas and adjacent areas. The extent of the hazardous location usually depends on area confinement and ventilation.

Where the spraying operation is located in a booth or room, the entire area of the booth or room is classified as a Class I or II, Division 1 location (the class depends on the material being sprayed). This classification is necessary because it is assumed that hazardous concentrations of gases (or vapor or dust) will be present in the area under normal

Figure 3-12. *An enclosed paint spray booth used in a fleet maintenance shop. This booth is illuminated through wired glass panels by fixtures mounted on the exterior of the booth.*

operating conditions. Where spraying is performed in open general areas, a Division 1 location extends for 20 ft (6.1 m) in all directions from the spray gun.

The space within 5 ft (1.52 m) in all directions from an open face or front of a spray booth is considered to be a Division 2 location if the spraying system is interlocked with the ventilation system. If the ventilation system is not interlocked with the spraying equipment, the Division 2 location extends 10 ft (3.05 m) from the open face. The space around other booth openings, such as conveyor openings, is a Class I or II, Division 2 location. (See Figure 3-13.)

When spraying is performed in an enclosed booth or room, the interior is still classified as a Division 1 location. The area within 3 ft

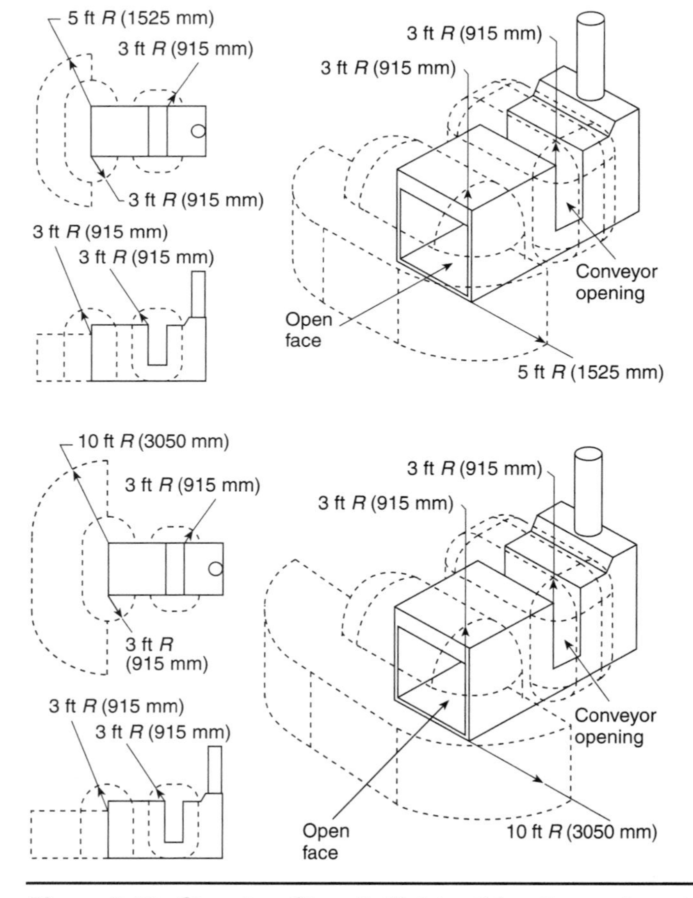

Figure 3-13. *Class I or Class II, Division 2 locations adjacent to a closed-top, open-face, or open-front spray booth.*

(914 mm) of the exterior of any door or other normally closed opening is classified as a Division 2 location. (See Figure 3-14.)

Open spraying processes generally result in a larger classified area. This is because vapors are not confined. Therefore, a larger area may be subject to overspray vapors under normal operating conditions and during a ventilation system failure. The spray area is classified as a Class I or II, Division 1 hazardous location. It will vary in size according to the arrangement of equipment and method of operation. The exact dimensions of the Division 1 location require a review of the installation by the authority having jurisdiction. A Division 2 location extends for 20 ft (6.1 m) horizontally beyond the Division 1 location to a height 10 ft (3.05 m) above the floor. (See Figure 3-15.)

Area classification is not necessary for outdoor spraying operations, such as the painting of buildings or bridges. Operations such as this type do not create excessive overspray deposits, and vapor-air mixtures are limited.

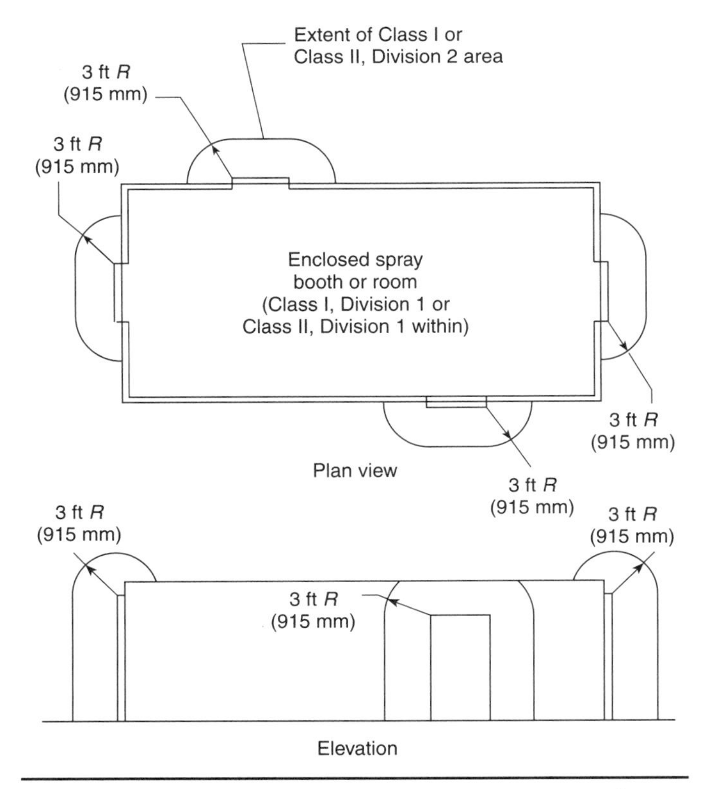

Figure 3-14. *Class I or Class II, Division 2 locations adjacent to openings in an enclosed spray booth or room.*

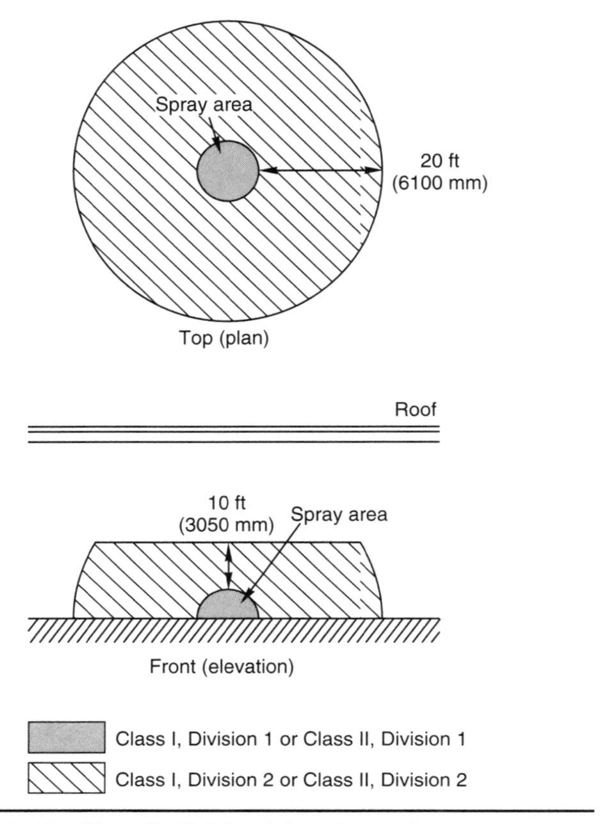

Figure 3-15. *Class II, Division 2 locations adjacent to an unenclosed spray operation.*

Occasionally the interior of the building may be spray painted as a building maintenance function. Such operations will not require classification; however, safeguards are necessary. Adequate ventilation should be provided during spraying and open flames must not be allowed. Containers should be kept tightly closed, and paint-soaked rags and other trash should be properly disposed of each day.

Combustible Overspray Deposits: Overspray deposits are a spray application problem, regardless of the type of spray application confinement system used. If allowed to build up, these deposits may present operational problems and a fire hazard.

Electrostatic systems produce the least amount of overspray, because charged particles are attracted to the charged workpiece. Confinement techniques, such as installation of paint spray booths, limit the overspray problem. Some booths are equipped with various types of filters to remove solid particulates. Booths may be designed with water-wash or oil-wash systems instead of a filter-type removal system. Open-floor spraying can present a difficult cleanup problem. The spray usually must be removed manually. It may also be necessary to use solvents to remove spray residue from the surrounding areas. NFPA 33 requires the use of solvents with flash points greater than 100°F (38°C). However, for cleaning spray nozzles and auxiliary equipment, a solvent with a lower flash point may be used if it does not have a flash point lower than the flash points of the spray coating materials. This permits the solvent used to dilute the paint to be used to clean the spraying equipment. These lower-flash-point solvents should be used in the spray booth or area only when the ventilation system is operating. Low-flash-point solvents should not be used outside of the spray area, since these areas usually have electrical equipment that is not suitable for use in a Division 2 or a Division 1 location.

Portable equipment lamps and other portable electric utilization equipment are generally prohibited from spray-painting areas. Portable equipment may accumulate spray deposits, hindering its operation and causing overheating. The preferred lighting method utilizes fixtures mounted on the exterior of the booth or room, which illuminate the booth through wired or tempered glass panels on the walls and/or ceilings. There may be instances where it may be necessary to illuminate spray areas not easily illuminated by the fixed lighting. For these applications, it is permissible to use portable lamps, but they must be of a type approved for Class I or II, Division 1 locations (whichever is applicable) if there are readily ignitible residues present.

There are special requirements for portable electric drying equipment, which is often used in automotive body shops. The dryer equipment must not operate during any spraying. A timed interlock circuit is required to allow a 3-minute purge of the booth before the dryer can be energized. The interlock must also deenergize the dryer during a failure of the ventilation system. It is not sufficient to depend on the interlocks for protection of drying equipment. The equipment must never be permitted to be located in the booth during spraying operations, because combustible overspray residue may accumulate on the equipment. Subsequent operation of the dryer may ignite these deposits.

Static Electricity: Static electricity can be a significant problem in electrostatic as well as nonelectrostatic air and airless spraying systems. Means must be provided to prevent static charges from reaching dangerous potentials. The problem can be especially severe with electrostatic systems, since the coating material is intentionally charged.

Most electrostatic systems employ a charging circuit, which limits the energy of a discharge arc to some value below the minimum value necessary for ignition when the safe sparking distance to the workpiece is maintained.

Electrostatic spraying systems should be equipped with interlocks that disable the power supply when the system is not in use. Additional safeguards may be required to prevent any grounded object from approaching closer than twice the safe sparking distance to the electrically charged high-voltage elements.

The likelihood of electrostatic ignition of electrostatic and nonelectrostatic systems can be minimized by bonding and grounding. Equipment to be bonded would include the fluid transfer system, the workpiece, the conveyor, and any conductive object within 10 ft (3.05 m).

Plastic parts may accumulate large electrostatic charges. If plastic parts are going to be located in an area where they are subject to static charges, the plastic should be impregnated with a conductive material to aid charge dissipation. For further information on static electricity, see Chapter 5.

Robotic Paint Spraying Equipment: Robots or manipulators are very common in large assembly operations, such as automobile assembly. They are usually used to operate welding equipment and paint spray guns. The majority of the paint spray robots have intrinsically safe electrical circuits. However, the spray gun is not intrinsically safe. If the robot wiring is intrinsically safe, the provisions of *NEC* Article 504 apply to the robot. However, Articles 501 and 502 will not normally apply to the robot. (See Figure 3-15.) General-purpose flexible cable of a material compatible with the sprayed material is permitted for intrinsically safe equipment. An explosionproof design would require the use of fixed-place equipment, partially fixed-place equipment, or equipment employing flexible connections. This may hinder the ability of the robot to move around the workpiece.

Intrinsically safe robots generally operate hydraulically. Servovalves are used to regulate the flow of hydraulic oil that changes the position of the robot's arm or manipulator. Resolvers and/or potentiometers may be used as feedback devices to the controller. The servovalve's resolvers and potentiometers are connected to safety barriers (associated intrinsically safe apparatus), which are located in the control cabinet. The barriers limit the energy present at each of these devices to a safe level. (See Section 4-2.1.2 for more information on intrinsically safe equipment.) If a barrier requires replacement, it is important to use the same model number. Barriers of the same basic voltage and current ratings may permit the robot to function, but they may not render the devices connected to them intrinsically safe.

Power equipment and safety barriers are normally installed in a control cabinet located outside the hazardous location, because the cabinet is usually unsuitable for hazardous location use. The hydraulic compressor is also usually located outside the booth or room, as few explosionproof hydraulic compressors are available. Cables and hydraulic hoses are routed through the wall of the booth to the robot.

Robots and spraying equipment often are separately listed by a testing laboratory. The end user should ascertain that the equipment is suitable for use with a spray robot. Some robots may be listed for nonhazardous location applications, such as welding. It is also possible that the paints or other solvents used may be incompatible with plastic materials used in robot construction.

3-1.2.7 Dipping and Coating Processes

Dipping and coating processes include a number of processes that use flammable or combustible materials, such as oil quenching, painting, impregnating, priming, and cleaning. Many of the coating or dipping materials are used above their flash points.

Dipping Processes: Dipping processes usually involve dipping workpieces in a tank of flammable or combustible liquid. A common application is to dip a hot metal object in a tank of quench oil to reduce the temperature of the object. Another common application is to dip an object in a tank of cleaning solvent to remove grease or other contaminants. However, high-flash-point solvents may also be used and are not ordinarily hazardous if they are not heated above their flash points.

When the tank contains a liquid being used at or above its flash point, the area is a Class I hazardous location. NFPA 34, *Standard for Dipping and Coating Processes Using Flammable or Combustible Liquids*[13], and *NEC* Article 516 provide requirements for area classification around dip tanks. (See Table 3-4.) Figure 3-16 illustrates the area classification of a typical dip tank. The dip tank and the drainboard are both considered to be vapor sources, because liquid may be on the drainboard flowing back toward the tank. Therefore, the limits of the hazardous location must be measured from these points. The area within a 5-ft (1.52-m) radius in all directions to the floor is a Class I, Division 1 location. In addition, pits within a 25-ft (7.62-m) radius are classified as Class I, Division 1 locations. This is because vapors accumulate in pits, where they are not readily dissipated. If the pit is located within 25 ft (7.62 m) but extends beyond the 25-ft (7.62-m) radius, the entire pit must be classified as Class I, Division 1, unless a vapor stop is provided.

The area between a 5-ft (1.52-m) and 8-ft (2.44-m) radius in all directions is classified as a Class I, Division 2 location. An additional Class I, Division 2 location extends within 3 ft (914 mm) of the floor for a distance of 25 ft (7.62 m) horizontally from the vapor source. This

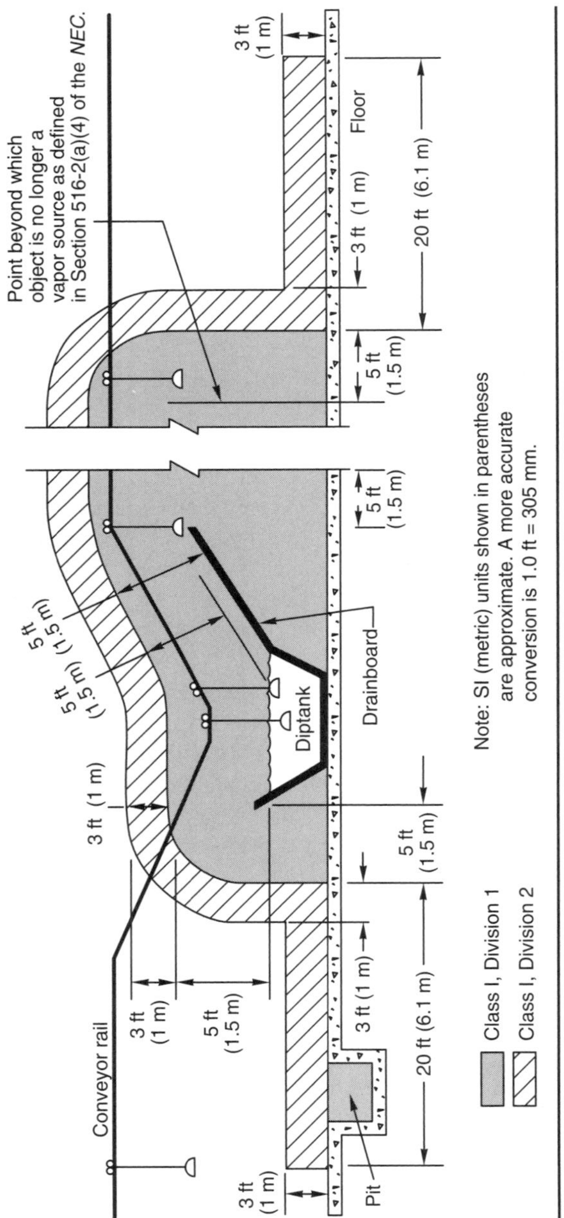

Figure 3-16. *The extent of Class I, Division 1 and Class I, Division 2 hazardous locations for open dipping processes.*

additional Division 2 location does not apply to tanks 5 ft^2 (0.465 m^2) or less and containing not more than 5 gal (18.93 L).

The classified area around a dip tank can be minimized by confining it to an enclosure, such as a room. However, the space within 3 ft (914 mm) of any opening in the enclosure is classified as a Class I, Division 2 location.

Dip tanks are often equipped with a cover held open by a wire or chain with a fusible link. If a fire occurs, the fusible link will melt, causing the cover to close. Fusible-link-actuated covers do not mitigate the need for area classification. A fire or explosion hazard may still be present. The link-operated cover only serves to extinguish a fire in the tank. (See Figure 3-17.)

Coating Processes: Common coating processes include roll coaters, flow coaters, and curtain coaters.

Roll coaters utilize one or more rollers covered with the coating material. The roller is in contact with a workpiece and usually an open pan of liquid coating or a nip formed by two rollers. The liquid coating may be supplied to the coating pan or nip from a larger tank or reservoir, or it may simply be poured into the pan.

Flow coating is a process in which a liquid coating is applied from

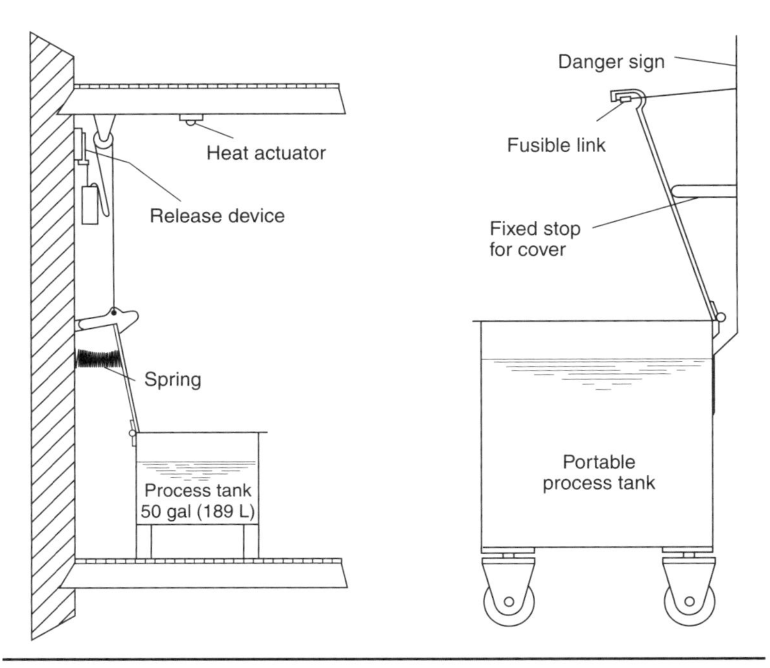

Figure 3-17. *Automatic-closing covers for small fixed and portable dip tanks.*

slots or nozzles to the workpiece. The coating is not atomized by the slots or nozzles. Therefore, excess coating is usually collected in a sump and recirculated to the reservoir. The workpiece is usually transported through the flow coater by a conveyor. The coater assembly usually includes an intermediate stage known as a drip tunnel. This tunnel allows excessive solvents to drip off the workpiece for recycling back to the process reservoir. At the beginning of its travel through the drip tunnel, the vapors around the workpiece may be readily ignitible. In the final stage, the workpiece travels through a drying oven.

Curtain coaters are used to coat flat or slightly curved items. They consist of a coating head that is supplied with coating from a reservoir. Coating material from the coating head overflows a wire installed on one side of the coating head. A continuous stream of coating material flows onto the workpiece. Excess coating collects in a sump and is returned to the reservoir. (See Figure 3-18.)

Coating processes are a potential explosion hazard where flammable liquids are used or where combustible liquids are used at or above their flash points. When either of these conditions exists, it is necessary to classify the area for electrical equipment. NFPA 34 provides the requirements for electrical area classification for coating processes.

Obviously, the area classification is the most severe in the immediate vicinity of the trough, sump, or tank below the workpiece. The area classification extends in all directions beyond this point. However, when workpieces move on a conveyor, the Division 1 and Division 2 locations extend a greater distance in the direction of conveyor movement. This is because the workpiece is usually wet and releasing vapors. A drainboard, which slopes toward the tank, is usually installed to prevent the liquids

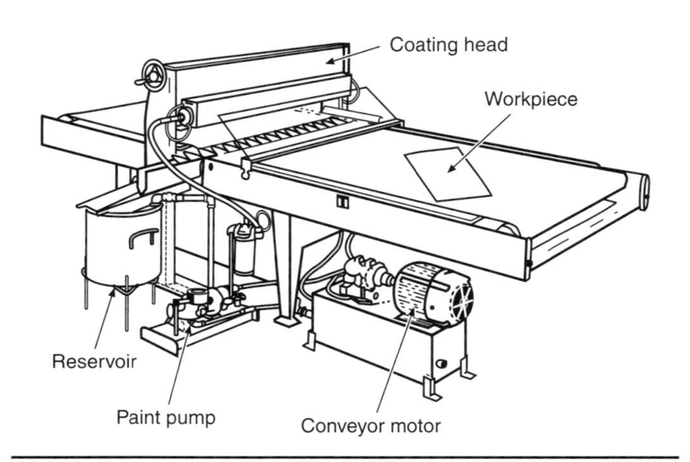

Figure 3-18. *Curtain coater.*

from being spilled on the floor and to minimize waste. Figure 3-19 illustrates the hazardous location around a typical process.

The principal advantage of the enclosed coating process is that it limits the extent of the hazardous location. NFPA 34 classifies the area adjacent to an enclosed process as nonhazardous with respect to the process. However, if there are openings in the enclosure, the space within 3 ft (914 mm) from any opening is classified as a Class I, Division 2 location. If a coating process is not enclosed, then it may be necessary to enforce the same classification as would be applied to an open process, such as a dip tank. The extent of the Division 2 classification need not be so severe when the vapor source has an area of 5 ft^2 (0.46 m^2) or less and when the open tank contents do not exceed 5 gal (18.93 L). However, in order to take advantage of this provision, the vapor concentration must not exceed 25 percent of the lower flammable limit (LFL) outside of the Division 1 location.

Processes are normally designed so that the drainboard covers most of the area where significant dripping from the workpiece occurs. The workpiece then travels through an intermediate portion of the drip tunnel, which is classified as a Division 2 location.

3-1.2.8 Ovens and Dryers Processing Flammable Materials

In most coating, dipping, and spray application processes, the workpieces are sent through a drying or curing oven to drive off any remaining

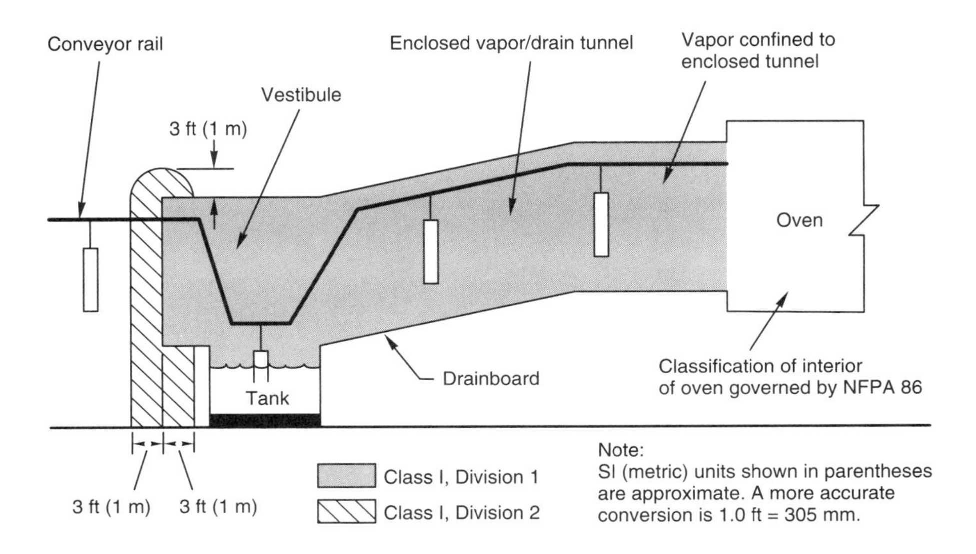

Figure 3-19. *The extent of Class I, Division 1 and Class I, Division 2 hazardous locations for enclosed coating processes.*

vapor. The interior of the oven is considered to be a nonhazardous location, because most of the vapor has been driven off by air drying before the workpiece enters the oven. Any remaining vapor is kept below the lower flammable limit (LFL) by ventilation in the oven. The ventilation is designed to keep any vapors below 25 percent of the LFL. When vapor concentrations may exceed 25 percent of the LFL, the oven may employ an inert atmosphere to minimize the hazard. During shutdown periods, vapors can accumulate at low points in the process oven. The ovens are usually designed to require a timed purge of the interior of the oven prior to operation. The ventilation of the oven and the purge prior to operation, as well as the inherent ignition source presented by most heating systems, negate the need for area classification in and around most properly designed process ovens. The design of process ovens is beyond the scope of this handbook. Further information can be found in NFPA 86, *Standard for Ovens and Furnaces*[14], and NFPA 86C, *Standard for Industrial Furnaces Using a Special Processing Atmosphere*[15].

3-1.2.9 Facilities for the Use, Processing, and Storage of Compressed Gas Cylinders

Pressurized cylinders are frequently used to facilitate storage and transportation of small amounts of flammable gases. The contents can be stored in a gaseous or liquid state. Gas cylinders are commonly used to store ordinary laboratory gases as well as some specialty laboratory gases. Small pressurized tanks can also be used to power small vehicles such as forklifts. They have also been used in some medium-sized, over-the-road cars, trucks, and buses.

Laboratories: The use of gases as well as flammable liquids in most laboratories is governed by NFPA 45, *Standard on Fire Protection for Laboratories Using Chemicals*[16]. The following laboratories are excluded:

1. Laboratories that are covered by Chapter 7 of NFPA 99, *Standard for Health Care Facilities*[17]
2. Laboratories that are pilot plants
3. Laboratories that are principally manufacturing plants
4. Laboratories that are incidental testing facilities
5. Physical, electronic, instrument, or similar laboratories that use small quantities of chemicals for incidental purposes such as cleaning

The electrical requirements of NFPA 45 are significant, since they modify the classification requirements of Article 500 of the *National*

Electrical Code. NFPA 45 requires all equipment to be installed in accordance with the *NEC*. However, it states that laboratory work areas and laboratory units are unclassified electrically. There are some exceptions to the rule. If extraordinarily hazardous conditions exist, it may be necessary to classify a laboratory work area or a part therein as a hazardous location. (See Table 3-5.)

A laboratory unit is defined as an enclosed space used for experiments or tests. A laboratory unit may or may not include offices, lavatories, and other contiguous rooms maintained for, or used by, laboratory personnel, and corridors within the unit. It may contain one or more separate laboratory work areas, or it may be an entire building.

A laboratory work area is defined as a room or space for testing, analysis, research, instruction, or similar activities that may involve the use of chemicals. It may or may not be enclosed.

Obviously, a laboratory work area or unit must comply with the other requirements of NFPA 45 to be permitted to be unclassified electrically. Important requirements include ventilation, fire protection, and the removal of cylinders that are not required for current laboratory projects.

Refrigerators, freezers, and other cooling equipment in a laboratory used to store flammable liquids must be suitable for the purpose. If there is electrical equipment within the outer shell, the equipment must meet the requirements for Class I, Division 1 locations. Equipment installed on the outside of the storage compartment must be either suitable for Class I, Division 2 locations, installed above the storage compartment,

Table 3-5 Class I Locations for Laboratories Within the Scope of NFPA 45

Location	Classification	Extent of Location
Laboratory	Nonhazardous	Applies to laboratory units or laboratory work areas
	Class I, Division 2	Interiors of refrigerator storage compartments used to store flammable liquids
	Class I, Division 1	Within the outer shell of a refrigerator storage compartment used to store flammable liquids (see text regarding equipment installation for exterior of storage compartment)
	Class I, Division 2	5 ft (1.5 m) to 8 ft (2.5 m) in all directions from vapor source or wetted surface, extending to the floor. 25 ft (7.5 m) horizontally and for 3 ft (1 m) above the floor
Laboratory (extraordinary hazard)		The extent and degree of the hazardous location is determined by the authority having jurisdiction

or located at a point on the outside surface of the compartment where the exposure will be minimal. If the laboratory work area or unit in which the refrigerator is installed has been judged to be an extraordinary hazard requiring classification, then the refrigerator must be suitable for that hazard, that is, approved for Class I, Division 1 or Division 2, as applicable.

Industrial Facilities Using Compressed Gas Cylinders: The most common usage of compressed gas cylinders is for the storage of liquefied petroleum gases (LP-Gas or LPG). These gases are made from crude oil and natural gas. The most frequently used liquefied petroleum gases are propane and butane. Propane is the most frequently used of the LP-Gases. It is used for cooking, heating, fuel for vehicles, cutting and welding, and so forth. Butane has some uses that are similar to propane. There are also applications in which a mixture of the two is used.

When LP-Gas is compressed in a container, it is liquefied within the container. A full container usually contains about 80 percent liquid; the remainder is in a gaseous state. When the liquid is vaporized, one volume of liquid equals about 270 volumes of gas. Therefore, a release from a tank can produce a substantial amount of vapor. The pressure within the container will vary with temperature.

Since there is a substantial leakage potential, NFPA 58, *Liquefied Petroleum Gas Code*[18], generally requires LP-Gas containers to be stored outside of buildings. Some exceptions are permitted for LP-Gas-fueled vehicles, industrial trucks, and stationary or portable engine fuel systems. These installations as well as some limited applications of fixed small tanks and portable containers must meet the provisions of NFPA 58.

The preferable storage location is outside or in a cutoff building of suitable construction. Where LP-Gas is to be distributed in a building, the cylinders are usually located outside of the building, with the gas cylinders connected to a manifold. Specific construction requirements can be found in NFPA 58.

Tanks may be refilled from a tank truck, or they may be brought to a facility where the LP-Gas is transferred from a storage tank into the smaller container. This transfer operation must be performed outdoors, because the risk of leakage is greatest during transfer operations.

The electrical classification requirements for LP-Gas locations are shown in Table 3-6. Figure 3-20 illustrates the extent of classification around a typical gas connection or pressure relief valve. These classified areas as specified in Table 3-6 vary from 497. The guidelines provided in NFPA 497 are based on a broad range of gases and vapors. The requirements shown in Table 3-7 are extracted from NFPA 58 and are based strictly on the use of compressed gases.

Table 3-6 Electrical Classification Requirements for LP-Gas Locations

Part	Location	Extent of Classified Area*	Equipment Shall Be Suitable for *National Electrical Code,* Class 1, Group D**
A	Unrefrigerated containers other than DOT cylinders and ASME vertical containers of less than 1000-lb water capacity	Within 15 ft in all directions from connections, except connections otherwise covered in Table 3-6	Division 2
B	Refrigerated storage containers	Within 15 ft in all directions from connections otherwise covered in Table 3-6	Division 2
		Area inside dike to the level of the top of the dike	Division 2
C	Tank vehicle and tank car loading and unloading***	Within 5 ft in all directions from connections regularly made or disconnected for product transfer	Division 1
		Beyond 5 ft but within 15 ft in all directions from a point where connections are regularly made or disconnected and within the cylindrical volume between the horizontal equator of the sphere and grade	Division 2
D	Gauge vent openings other than those on DOT cylinders and ASME vertical containers of less than 1000-lb water capacity	Within 5 ft in all directions from point of discharge	Division 1
		Beyond 5 ft but within 15 ft in all directions from point of discharge	Division 2
E	Relief device discharge other than those on DOT cylinders and ASME vertical containers of less than 1000-lb water capacity, and vaporizers	Within direct path of discharge	Division 1 *Note: Fixed electrical equipment preferably should not be installed.*
F	Pumps, vapor compressors, gas-air mixers, and vaporizers (other than direct-fired or indirect-fired with an attached or adjacent gas-fired heat source)		
	Indoors without ventilation	Entire room and any adjacent room not separated by a gastight partition	Division 1
		Within 15 ft of the exterior side of any exterior wall or roof that is not vaportight or within 15 ft of any exterior opening	Division 2
	Indoors with adequate ventilation****	Entire room and any adjacent room not separated by a gastight partition	Division 2

Table 3-6 ***(continued)***

Part	Location	Extent of Classified Area*	Equipment Shall Be Suitable for *National Electrical Code,* Class 1, Group D**
	Outdoors in open air at or above grade	Within 15 ft in all directions from this equipment and within the cylindrical volume between the horizontal equator of the sphere and grade	Division 2
G	Vehicle fuel dispenser	Entire space within dispenser enclosure, and 18 in. horizontally from enclosure exterior up to an elevation 4 ft above dispenser base. Entire pit or open space beneath dispenser	Division 1
		Up to 18 in. aboveground within 20 ft horizontally from any edge of enclosure *Note: For pits within this area, see Part H of this table.*	Division 2
H	Pits or trenches containing or located beneath LP-Gas valves, pumps, vapor compressors, regulators, and similar equipment		
	Without mechanical ventilation	Entire pit or trench	Division 1
		Entire room and any adjacent room not separated by a gastight partition	Division 2
		Within 15 ft in all directions from pit or trench when located outdoors	Division 2
	With adequate mechanical ventilation	Entire pit or trench	Division 2
		Entire room and any adjacent room not separated by a gastight partition	Division 2
		Within 15 ft in all directions from pit or trench when located outdoors	Division 2
I	Special buildings or rooms for storage of portable containers	Entire room	Division 2
J	Pipelines and connections containing operational bleeds, drips, vents, or drains	Within 5 ft in all directions from point of discharge	Division 1
		Beyond 5 ft from point of discharge, same as Part F of this table	

Table 3-6 ***(continued)***

Part	Location	Extent of Classified Area*	Equipment Shall Be Suitable for *National Electrical Code,* Class 1, Group D**
K	Container filling		
	Indoors with adequate ventilation****	Within 5 ft in all directions from a point of transfer	Division 1
		Beyond 5 ft and entire room	Division 2
	Outdoors in open air	Within 5 feet in all directions from a point of transfer	Division 1
		Beyond 5 ft but within 15 ft in all directions from point of transfer and within the cylindrical volume between the horizontal equator of the sphere and grade	Division 2

**The classified area must not extend beyond an unpierced wall, roof, or solid vaportight partition.*

***See Article 500 — "Hazardous Locations" — in NFPA 70 (ANSI) for definitions of classes, groups, and divisions.*

****When classifying extent of hazardous area, consideration must be given to possible variations in the spotting of tank cars and tank vehicles at the unloading points and the effect these variations in actual spotting point may have on the point of connection.*

*****Where specified for the prevention of fire or explosion during normal operation, ventilation is considered adequate where provided in accordance with the provisions of NFPA 58.*

For SI units: 18 in. = 256 mm; 4 ft = 1.2 m; 5 ft = 1.5 m; 15 ft = 5 m; 20 ft = 6 m.

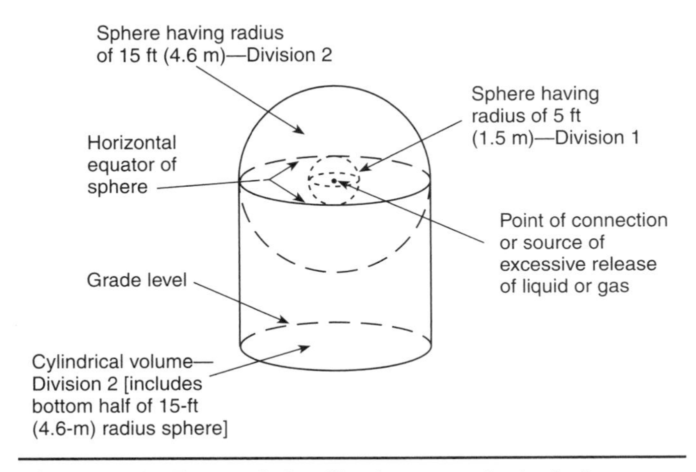

Figure 3-20. *Extent of classification around a typical gas connection or pressure relief valve.*

Table 3-7 Recommended Classification of Areas in Compressor Stations

Location	Class I, Group D	Extent of Location
	Natural Gas Areas	
Compressor building	Division 2	Basement
	Division 2	Operating floor
	Division 2	Exterior stairwell or pit
	Division 2	Electrical attachments to exterior
Odorant building	Division 2	Interior
	Division 2	Exterior stairwell or pit
	Division 2	Electrical attachments to exterior
Measurement and regulation building	Division 1	Interior
	Division 2	Control gas vented inside within 10 ft of exterior
	Division 2	Control gas vented outside above roof line
	Division 2	Exterior pit or trench
	Division 2	Electrical attachments to exterior
Dehydration building	Nonhazardous	
Telemetering building	Nonhazardous	
Auxiliary building	Nonhazardous	
Yard area	Division 1	5-ft radius from point source of blowoff or relief
	Division 2	Between 5-ft radius and 15-ft radius from point source of blowoff or relief
	Division 2	15-ft radius from valves, flanges, or screwed fittings
	Nonhazardous	General yard area
	Liquefied Petroleum Gas Areas	
Compresor building (for heavier-than-air gas)	Division 1	Basement
	Division 2	Operating floor
	Division 1	Exterior stairwell or pit
	Division 2	Electrical attachments to exterior
	See natural gas classification above	For lighter-than-air gas
	Synthetic Natural Gas Areas	
Liquid pump building	Division 1	Interior
	Division 2	Within 10 ft of exterior
	Division 1	Exterior pit or trench
	Division 2	Electrical attachments to exterior
Liquid process building	Division 1	Interior
	Division 2	Within 10 ft of exterior
	Division 1	Exterior pit or trench
	Division 2	Electrical attachments to exterior

Table 3-7 ***(continued)***

Location	Class I, Group D	Extent of Location
Refrigerated liquid storage tank (water capacity in excess of 1200 gallons)	Division 1	Interior
	Division 2	Within 10 ft of exterior
	Division 1	Above grade within retaining dike
	Division 1	Exterior pit or trench
	Division 2	Electrical attachments to exterior

Source: American Gas Association, XFO 277[19].

3-1.2.10 Facilities Processing Natural Gas

Natural gas is a mixture of naturally occurring gases. The principal component of natural gas is methane, but it also usually includes ethane, propane, and butane. It may occur as a result of several phenomena. It can be manufactured from decaying organic material, or it can be captured from oil-bearing strata. In order for a source of gas to be considered usable for gas-transmission purposes, the source must be of a considerable size. The most common sources of natural gas are gas wells, oil wells, and facilities where petroleum products are refined. Natural gas is a colorless, odorless gas. Ethyl mercaptan or some other odorant is added to natural gas to give it a distinctive odor, which aids in leak detection.

Natural gas well head production stations (or gas gathering stations) are usually gas compressor stations. Gas compressor stations are also located at storage facilities and at intermediate points along a transmission pipeline. A compressor station may consist of a compressor building, a pump room, odorant buildings, and gas turbine buildings. The gas compressor facilities compress the gas for transmission in the pipeline.

Many compressor stations are located in remote areas where electric utility power may not be available. These stations usually derive station power from the natural gas supply. The prime mover may be a reciprocating engine generator or a gas turbine. Since the fuel is inexpensive, compressor stations located in or near metropolitan areas will also use the natural gas supply as a primary or secondary power supply.

The American Gas Association has published XFO 277, a recommended practice on the classification of gas compressor stations. A summary of the recommended classification of areas can be found in Table 3-7.

Each room or building through which the gas piping is routed requires careful evaluation of the hazards present for electrical classification. The typical hazards in each of the areas include leakage of gas around seals, fittings and flanges, pumps, and pressure relief valves.

During periods of peak demand, gas utility companies often need to supplement their normal pipeline supply with a local supply. This is known as peak shaving. Usually a utility will use a supply of liquefied natural gas (LNG) or liquefied petroleum gas (LPG). Liquefied petroleum gas is usually predominantly propane or butane, with lesser amounts of ethane, ethylene, propylene, iso-butane, and butylene. In addition to peak shaving, utilities will often operate stand-alone LP-Gas distribution systems to supply a small subdivision on an interim basis. Table 3-8 covers electrical requirements for utility-owned LP-Gas facilities.

Special Considerations for Liquefied Natural Gas: Liquefied natural gas is typically supplied to gas distribution systems through pipelines. However, it may be shipped to a distribution center by truck, tank car, or tanker ship. The most practical state in which to ship large quantities of natural gas is in a liquefied state. In order to maintain natural gas in a liquid state, it must be stored at a temperature slightly above its boiling point temperature of -260°F (-162°C).

LNG is normally stored in large refrigerated outdoor tanks. Most compressor pump evaporators and other process equipment are also located outdoors. LNG is rarely used indoors in a liquefied state. (See Figure 3-21.)

Such extremely cold temperatures create some unusual problems. The temperature extremes cause many materials to lose all elasticity and become brittle. Electrical and mechanical equipment in contact with LNG must be suitable for operation at cryogenic temperatures. Means must also be provided to prevent exposure of incompatible equipment to cryogenic temperatures. Equipment that is directly exposed must be physically isolated from the incompatible equipment. Seals are provided in conduit and cable systems to minimize the passage of gases and vapors from one portion of the electrical system to another through the conduit system or through cable. Unless they are specifically designed for the purpose, seals are not usually designed to prevent the passage of gases or vapors across the seal at any temperature or pressure differential. Even at very low pressures, there may be migration of gas across the seal.

In LNG facilities, pressure differential across a seal can be particularly hazardous. Migration of supercooled LNG into a conduit system at a pressure above atmospheric may cause catastrophic failure of downstream conduit seals. This is due to the shock of the sudden temperature extreme plus the effect of the pressure. If failure of a primary seal would allow flammable liquids to enter the conduit, a secondary seal barrier or other means is required. Drains, ventilated fittings, or other means

Table 3-8 Electrical Requirements for Utility-Owned LP-Gas Facilities

Part	Extent of Classified Area*	Location	Class 1, Group D****
A	Within 15 ft (4.6 m) in all directions from connections, except connections otherwise covered in this table	Nonrefrigerated container	Division 2
B	Within 15 ft (4.6 m) in all directions from connections, except for connections otherwise covered in this table	Refrigerated	Division 2
	Area inside dike to a level of the top of the dike		Division 2
C	Within 5 ft (1.5 m) in all directions from connections regularly made or disconnected for product transfer	Tank vehicle and tank car unloading**	Division 1
	Beyond 5 ft (1.5 m) but within 15 ft (4.6 m) in all directions from a point where connections are regularly made or disconnected and with the cylindrical volume between the horizontal equator of the sphere and grade (See Figure 3-20.)		Division 2
D	Within 5 ft (1.5m) in all directions from point of discharge	Gauge vent	Division 1
	Beyond 5 ft (1.5 m) but within 15 ft (4.6 m) in all directions from point of discharge		Division 2
E	Within direct path of discharge	Relief valve discharge	Division 1 *Note: Fixed electrical equipment should preferably not be installed.*
	Within 5 ft (1.5 m) in all directions from point of discharge		Division 1
	Beyond 5 ft (1.5 m) but within 15 ft (4.6 m) in all directions from point of discharge except within a path of discharge		Division 2
F		Pumps, compressors, gas-air mixers, meter areas, calorimeters other than open flame types, and vaporizers other than direct fired	
	Entire room and any adjacent room not separated by a gastight partition*	Indoors without ventilation	Division 1

Table 3-8 ***(continued)***

Part	Extent of Classified Area*	Location	Class 1, Group D****
	Within 15 ft (4.6 m) of the exterior side of any exterior wall or roof that is not vaportight or within 15 ft (4.6 m) of any exterior opening		Division 2
	Entire room and any adjacent room not separated by a gastight partition*	Indoors with adequate ventilation***	Division 2
	Within 15 ft (4.6 m) in all directions from equipment and within the cylindrical volume between the horizontal equator of the sphere and grade (See Figure 3-20.)	Outdoors, at or above grade	Division 2
G		Pits or trenches containing equipment such as pumps, compressors (other than direct fired vaporizers), and similar equipment (also pits or trenches located beneath classified areas)	
	Entire pit or trench	Without mechanical ventilation	Division 1
	Entire room and any adjacent room not separated by a gastight partition when located indoors		Division 2
	Within 15 ft (4.6 m) in all directions from pit or trench when located outdoors		Division 2
	Entire pit or trench	With adequate mechanical ventilation***	Division 2
	Entire room and any adjacent room not separated by a gastight partition when located indoors		Division 2
	Within 15 ft (4.6 m) in all directions from pit or trench when located outdoors		Division 2
H	Within 5 ft (1.5 m) in all directions from point of discharge	Pipelines and connections containing operational bleeds, drips, vents, or drains	Division 1
	Beyond 5 ft (1.5 m) from point of discharge, same as Part F of this table		

Source: NFPA 59, Standard for the Production, Storage, and Handling of Liquefied Petroleum Gases at Utility Gas Plants[20].

**The classified area must not extend beyond an unpierced wall, roof, or solid vaportight partition.*

***When determining the extent of a classified area, consideration must be given to possible variations in the spotting of tank cars and tank vehicles at the unloading point and the effect that these variations of actual spotting point may have on the point of connection.*

****Ventilation is considered adequate when provided in accordance with the provisions of NFPA 59.*

*****See Article 500 — "Hazardous Locations" — NFPA 70, for definitions of classes, groups, and divisions.*

Figure 3-21. *LNG is usually stored in large outdoor refrigerated storage tanks.*

are usually provided between the primary and secondary seals to minimize the likelihood that LNG will travel into the conduit system. Leakage from these vents, drains, or other devices also will make it obvious that there is leakage through the primary seal.

Electrical equipment suitable for a Class I, Group D environment may be necessary if natural gas is or may be present in the air in explosive or ignitible concentrations. Complete requirements for LNG facilities can be found in NFPA 59A, *Standard for the Production, Storage, and Handling of Liquefied Natural Gas (LNG)*[21]. A summary of the electrical area classification requirements can be found in Table 3-9.

Static bonding protection is not normally required when LNG tank cars, tank vehicles, or container ships are unloaded. This is because transfer systems use metal pipe or flexible metal hoses. However, if a loading or unloading system utilizes cathodic protection, protective measures must be taken to prevent a current that could ignite the vapor.

Table 3-9 Electrical Area Classification Requirements for LNG

Part	Location	Group D, Division*	Extent of Classified Area**
A	LNG storage containers with vacuum breakers	2	Entire container interior
B	LNG storage container area		
	Indoors	1	Entire room
	Outdoor above-ground containers (other than small containers)*****	1	Open area between a high-type dike and container wall where dike wall height exceeds distance between dike and container walls (See Figure 3-22.)
		2	Within 15 ft (4.5 m) in all directions from container walls and roof, plus area inside a low-type diked or impounding area up to the height of the dike impoundment wall (See Figure 3-23.)
	Outdoor below-ground containers	1	Within any open space between container walls and surrounding grade or dike (See Figure 3-24.)
		2	Within 15 ft (4.5 m) in all directions from roof and sides (See Figure 3-24.)
C	Nonfired LNG process areas containing pumps, compressors, heat exchangers, pipelines, connections, small containers, and so on		
	Indoors with adequate ventilation***	2	Entire room and any adjacent room not separated by a gastight partition and 15 ft (4.5 m) beyond any wall or roof ventilation discharge vent or louver
	Outdoors in open air at or above grade	2	Within 15 ft (4.5 m) in all directions from this equipment and within the cylindrical volume between the horizontal equator of the shpere and grade (See Figure 3-20.)
D	Pits, trenches, or sumps located in or adjacent to Division 1 or 2 areas	1	Entire pit, trench, or sump
E	Discharge from relief valves	1	Within direct path of relief valve discharge
F	Operational bleeds, drips, vents, or drains		
	Indoors with adequate ventilation***	1	Within 5 ft (1.5 m) in all directions from point of discharge
		2	Beyond 5 ft (1.5 m) and entire room and 15 ft (4.5 m) beyond any wall or roof ventilation discharge vent or louver
	Outdoors in open air at or above grade	1	Within 5 ft (1.5 m) in all directions from point of discharge

Table 3-9 *(continued)*

Part	Location	Group D, Division*	Extent of Classified Area**
		2	Beyond 5 ft (1.5 m) but within 15 ft (4.5 m) in all directions from point of discharge
G	Tank car, tank vehicle, and container loading and unloading****		
	Indoors with adequate ventilation***	1	Within 5 ft (1.5 m) in all directions from connections regularly made or disconnected for product transfer
		2	Beyond 5 ft (1.5 m) and entire room and 15 ft (4.5 m) beyond any wall or roof ventilation discharge vent or louver
	Outdoors in open air at or above grade	1	Within 5 ft (1.5 m) in all directions from connections regularly made or disconnected for product transfer
		2	Beyond 5 ft (1.5 m) but within 15 ft (4.5 m) in all directions from a point where connections are regularly made or disconnected, and within the cylindrical volume between the horizontal equator of the sphere and grade (See Figure 3-20.)
H	Electrical seals and vents specified in Sections 7-6.3, 7-6.4, and 7-6.5 of NFPA 59A	2	Within 15 ft (4.5 m) in all directions from the equipment and within the cylindrical volume between the horizontal equator of the sphere and grade

**See Article 500 — "Hazardous (Classified) Locations" — in NFPA 70 (ANSI) for definitions of classes, groups, and divisions. Most of the flammable vapors and gases found within the facilities covered by NFPA 59A are classified as Group D. Ethylene is Group C. Much available electrical equipment for hazardous locations is suitable for both groups.*

***The classified area must not extend beyond an unpierced wall, roof, or solid vaportight partition.*

****Ventilation is considered adequate when provided in accordance with the provisions of NFPA 59A.*

*****When classifying the extent of a hazardous area, consideration must be given to possible variations in the spotting of tank cars and tank vehicles at the unloading points and the effect these variations of actual spotting points may have on the point of connection.*

******Small containers are those that are portable and of less than 200 gal (760 L) capacity.*

Large metal tanks do not normally require air terminals for lightning protection. However, if LNG tanks are installed on nonconductive foundations, the tanks should be connected to a grounding electrode or grounding electrode system. The grounding electrode will minimize the likelihood of a difference of potential between the tank and ground. In addition, it will protect the foundation of the tank from lightning damage.

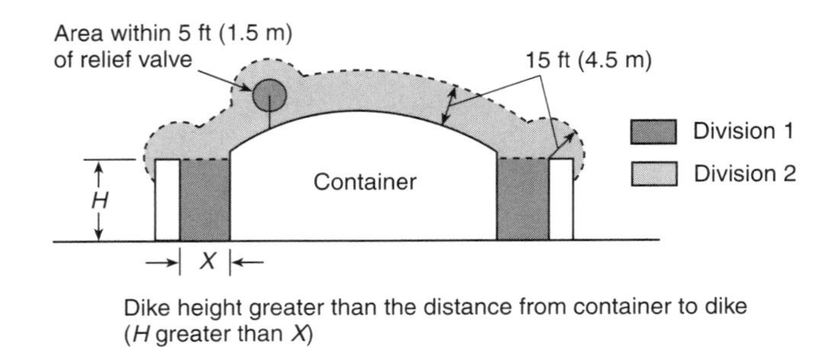

Figure 3-22. *Dike height less than distance from container dike (H greater than X).*

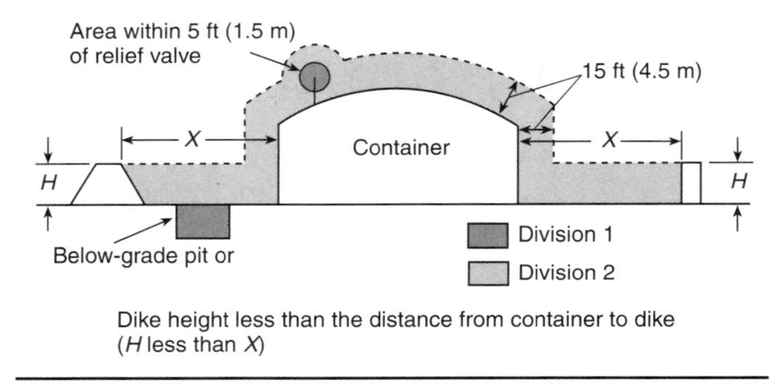

Figure 3-23. *Dike height less than distance from container to dike (H less than X).*

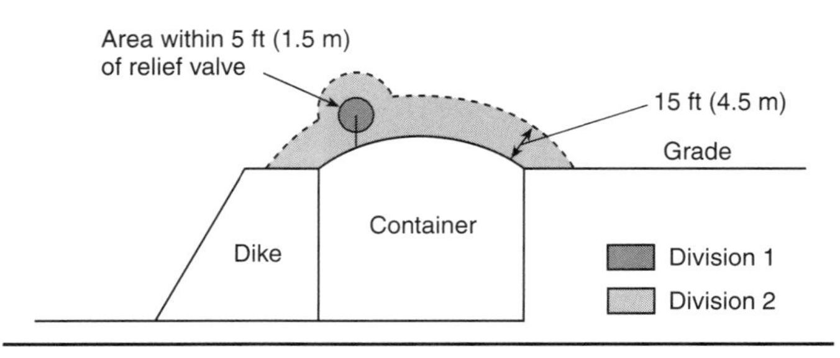

Figure 3-24. *Container with liquid level below grade or top of dike.*

3-1.2.11 General Storage and Dispensing of Flammable Liquids

Chapter 4 of NFPA 30 covers requirements for the general storage of flammable liquids in drums and other containers not exceeding 60 gal (227 L), and portable tanks of 60- to 660-gal (227- to 2498-L) capacity.

Many industrial facilities have need for small amounts of flammable or combustible solvents for mixing with other materials, cleaning, degreasing, and so on. These solvents are sold in bottles, cans, and drums. Glass bottles are often used for laboratory-grade materials. Flammable-liquid bottles are usually available in sizes of 1 gal (3.79 L) and less. Cans are usually available in sizes of 1 to 5 gal (3.79 to 18.95 L). The typical flammable liquid drum currently in use in the United States is a drum with a capacity of 55 gal (208 L). Smaller containers of 1 and 5 gal (3.79 and 18.95 L) are also common. Smaller amounts of fluid are dispensed into 1-to 5-gal (3.79-to 18.95-L) cans for use in the plant.

Where flammable liquid drums are stored indoors, they are normally stored in a ventilated room of noncombustible construction. Ventilation at a rate of 1 ft^3/minute · ft^2 (0.028 m^3/minute · 0.093 m^2) of floor area [minimum flow rate of 150 ft^3/minute (4.2 m^3/minute)] is usually provided in order to prevent the accumulation of vapor. The floor is usually diked and equipped with drains in order to prevent the release of any spilled liquid into main plant areas. Where the room is used for storage only, any electrical equipment in the room must be suitable for Class I, Division 2 locations. The group will depend on the materials stored in the room. The selection of equipment should be based on the group of the worst-case material stored in the room, Group A being the most severe and Group D being the least severe. A listing of the group classifications of several common materials can be found in Appendix A, Table A-4, and NFPA 497.

Where Class II or III liquids are stored below their flash points, it is not necessary to use equipment suitable for hazardous locations, unless there are also some Class IA or Class IB materials present. This classification assumes that the room is used primarily for storage. For rooms or buildings of 1000 ft^2 (93 m^2) and less, a limited amount of dispensing is permitted. In the area around a dispensing operation or drum filling station, a Class I, Division 1 location exists for a 3-ft (914-mm) radius from the vent pipe and the dispensing faucet.

Drum filling stations may be located indoors or outdoors. When located indoors, an adequate source of ventilation should be provided to minimize the extent of the hazardous location. Generally, a source of 1 ft^3/minute · ft^2 (0.028 m^3/minute · 0.093 m^2) of floor area [minimum of 150 ft^3/minute (4.2 m/minute)] should be sufficient. A hazardous location will still exist around the fill pipe and the vent of the drum, because vapors will be liberated. Figure 3-25 illustrates the classification recommended by NFPA 497. Without adequate ventilation, the hazardous area could be much larger.

For information on the classification of laboratories, see the discussion of laboratories in Section 3-1.2.9.

The dispensing of liquids from drums can generate static electricity. Static charge buildup may be minimized by bonding the drum to the

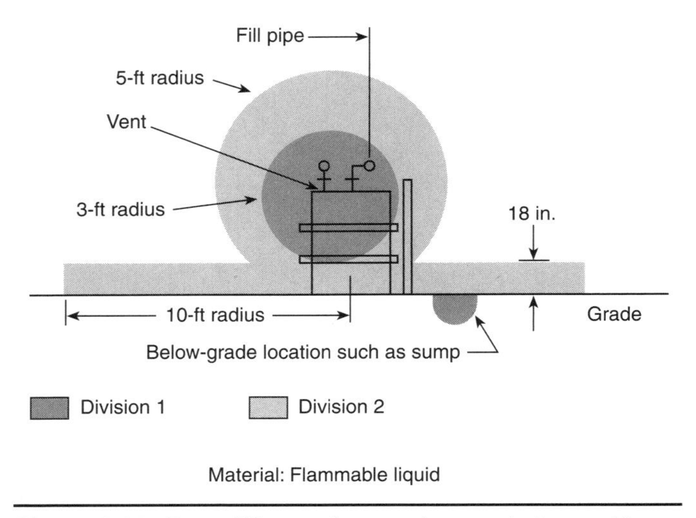

Figure 3-25. *Drum filling station, outdoors or indoors with adequate ventilation.*

frame of the dispensing rack and grounding the rack. (See Section 5-1.) Each container into which Class IA or Class IB liquids are to be dispensed should be bonded to the metal drum from which the fluid is being dispensed.

Flammable liquid storage rooms are often very corrosive environments. Static bonding wires and clamps should be inspected periodically to assure that they are still electrically continuous.

3-1.2.12 Drycleaning Plants

Drycleaning plants or systems are classified into six general types. The type number of the plant or system is based on the types of solvents used:

Type I — Systems employing Class I solvents [flash point less than 100°F (38°C)], such as naphtha [flash point 50°F (10°C)].

Type II — Systems employing Class II solvents [flash point 100°F (38°C) to 140°F (60°C)].

Type IIIA — Systems employing Class IIIA solvents [flash point 140°F (60°C) to 200°F (93.4°C)].

Type IIIB — Systems employing Class IIIB solvents [flash point over 200°F (93.4°C)]. These liquids are usually specially compounded oils.

Type IV — Systems that use solvents that have been classified as nonflammable (Class IV solvents). These plants are facilities in which dry cleaning is not conducted by the public.

Type V — Systems employing solvents that have been classified as nonflammable (Class IV solvents). These facilities are dry cleaning stores in which the public can use coin-operated drycleaning equipment.

NFPA 32, *Standard for Drycleaning Plants*[22], has prohibited construction of Type I plants or systems for several years. The only type of drycleaning plant that requires classification as a hazardous location is a Type II plant. These plants are rarely built, due to the restrictions on construction, building use, equipment requirements, and so forth. NFPA 32 classifies the entire Type II dry cleaning room as a Class I, Division 2 location. The group depends on the solvent used.

The generation of static electricity can be significant in drycleaning operations. To minimize the static buildup between equipment, all non-current-carrying metal parts of storage tanks, treatment tanks, filters, pumps, piping ductwork, drycleaning units, stills, drying cabinets, tumblers, and other equipment in the dry cleaning room must be bonded together and grounded. (See Section 5-1.) Isolated equipment must also be grounded.

3-1.2.13 Bulk Storage Plants

Bulk storage plants are facilities where gasoline or other volatile flammable liquids are stored in tanks having an aggregate capacity of one carload or more. (See Figure 3-26.) A typical plant may have several large tanks arranged together in a common area or "tank farm." The tanks may be individually diked to contain a spill from any tank. As an

Figure 3-26. *A bulk storage plant with individual diked tanks.*

alternative, the tanks may be located within a larger common diked area with runoff to an impounding area large enough to contain all of the liquid from the largest tank. The only equipment that is usually found within the diked area of the tank farm is piping and valves. A bulk storage facility may also include an office building and other buildings for maintenance, pumps, vehicle garages, and so forth.

Each bulk storage plant usually has loading and unloading facilities. In coastal and major waterway areas, a marine terminal may serve to supply the tank farm. In inland areas, the loading and unloading may be via railroad tank cars. Local distribution of fuel is usually via a motor tank truck terminal.

Loading and unloading of fluids is usually performed at high flow rates. In tank vehicles with open domes, the high transfer rate can result in the generation of static electricity. The charged fluid seeks to relax its charge upon contact with the grounded metal of the tank. The resulting static discharge can cause ignition of Class I flammable liquids. However, the hazard can also exist if Class II or III liquids are being dispensed into a tank that previously contained Class I flammable liquids. The residual vapors can be ignited by static discharges from the newly introduced fluid. Static charges can be kept at safe levels with proper bonding. Means must be provided to bond the open-domed vehicle to the fill pipe to minimize the buildup of static charge through the fluid. For further information on static electricity, see Section 5-1.

Vapor is often lost into the atmosphere during loading and unloading operations. Vapor recovery units are provided to recapture the lost vapor. Since ignitible quantities of vapor may exist in the absorption or adsorption tanks, all or part of the unit are often considered to be located in a hazardous location.

A bulk storage plant may have a number of hazardous (classified) locations. NFPA 30 and Article 515 of the *National Electrical Code* specify the electrical area classification scheme for bulk storage plants. (See Table 3-10.)

Marine terminals handle large amounts of fluid at high flow rates. Transfers of this magnitude require special classification consideration. Figure 3-27 illustrates the minimum area classification specified by NFPA 30 for marine terminal facilities. This classification applies to marine terminals that handle flammable liquids. (See Figure 3-28.) It does not apply to terminals that process LP-Gas or LNG. The extent of the Division 2 hazardous locations in marine terminals is very large due to the large volume of fluid present and the high flow rates.

Salt water corrosion can be a significant problem in marine terminal facilities. It can affect equipment, conductors, and bonding jumpers. Maintenance and inspection programs are vital to ensure that corrosion does not create a dangerous condition.

Table 3-10 Electrical Equipment Classified Areas — Bulk Plants

Location	*NEC* Class I, Group D Division	Extent of Classified Area
Tank vehicle and tank car*		
Loading through open dome	1	Within 3 ft of edge of dome, extending in all directions
	2	Area between 3 ft and 15 ft from edge of dome, extending in all directions
Loading through bottom connections with atmospheric venting	1	Within 3 ft of point of venting to atmosphere, extending in all directions
	2	Area between 3 ft and 15 ft from point of venting to atmosphere, extending in all directions. Also, up to 18 in. above grade within a horizontal radius of 10 ft from point of loading connection.
Loading through closed dome with atmospheric venting	1	Within 3 ft of open end of vent, extending in all directions
	2	Area between 3 ft and 15 ft from open end of vent, extending in all directions. Also, within 3 ft of edge of dome, extending in all directions
Loading through closed dome with vapor control	2	Within 3 ft of point of connection of both fill and vapor lines, extending in all directions
Bottom loading with vapor control; any bottom unloading	2	Within 3 ft of point of connections extending in all directions. Also, up to 18 in. above grade within a horizontal radius of 10 ft from point of connection.
Drum and container filling outdoors, or indoors with adequate ventilation	1	Within 3 ft of vent and fill opening, extending in all directions
	2	Area between 3 ft and 5 ft from vent or fill opening, extending in all directions. Also, up to 18 in. above floor or grade level within a horizontal radius of 10 ft from vent or fill opening.
Tank — above ground**		
Shell, ends, or roof and dike area	2	Within 10 ft from shell, ends, or roof of tank. Area inside dikes to level of top of dike.
Vent	1	Within 5 ft of open end of vent, extending in all directions
	2	Area between 5 ft and 10 ft from open end of vent, extending in all directions
Floating roof	1	Area above the roof and within the shell
Pits		
Without mechanical ventilation	1	Entire area within pit if any part is within a Division 1 or 2 classified area
With mechanical ventilation	2	Entire area within pit if any part is within a Division 1 or 2 classified area
Containing valves, fittings, or piping, and not within a Division 1 or 2 classified area	2	Entire pit

Table 3-10 ***(continued)***

Location	*NEC* Class I, Group D Division	Extent of Classified Area
Pumps, bleeders, withdrawal fittings, meters, and similar devices		
Indoors	2	Within 5 ft of any edge of such devices, extending in all directions. Also, up to 3 ft above floor or grade level within 25 ft horizontally from any edge of such devices.
Outdoors	2	Within 3 ft of any edge of such devices, extending in all directions. Also, up to 18 in. above grade level within 10 ft horizontally from any edge of such devices.
Storage and repair garage for tank vehicles	1	All pits or spaces below floor level
	2	Area up to 18 in. above floor or grade level for entire storage or repair garage
Drainage ditches, separators, impounding basins	2	Area up to 18 in. above ditch, separator, or basin. Also, up to 18 in. above grade within 15 ft horizontally from any edge.
Garage for other than tank vehicles	Ordinary	If there is any opening to these rooms within the extent of an outdoor classified area, the entire room shall be classified the same as the area classification at the point of the opening
Outdoor drum storage	Ordinary	
Indoor warehousing where there is no flammable liquid transfer	Ordinary	If there is any opening to these rooms within the extent of an indoor classified area, the room shall be classified the same as if the wall, curb, or partition did not exist
Office and rest rooms	Ordinary	

Source: NFPA 30.

**When classifying extent of area, consideration shall be given to the fact that tank cars or tank vehicles may be spotted at varying points. Therefore, the extremities of the loading or unloading positions shall be used.*

***For Tanks — Underground, see Chapter 5 of NFPA 30A.*

For SI units: 1 in. = 2.5 cm; 1 ft = 0.30 m.

3-1.2.14 Battery Charging Areas

Batteries are used in an increasing number of industrial applications. For many years, they have been used to power electric trucks and to provide power for protective relays. They are gaining increasing use in uninterruptable power supplies for communications, data processing, alarm, and industrial process equipment.

Batteries generate a small amount of hydrogen and other gases while they are being charged or discharged. Hydrogen is a lighter-than-air gas that is readily ignitible. Article 480 of the *NEC* governs stationary

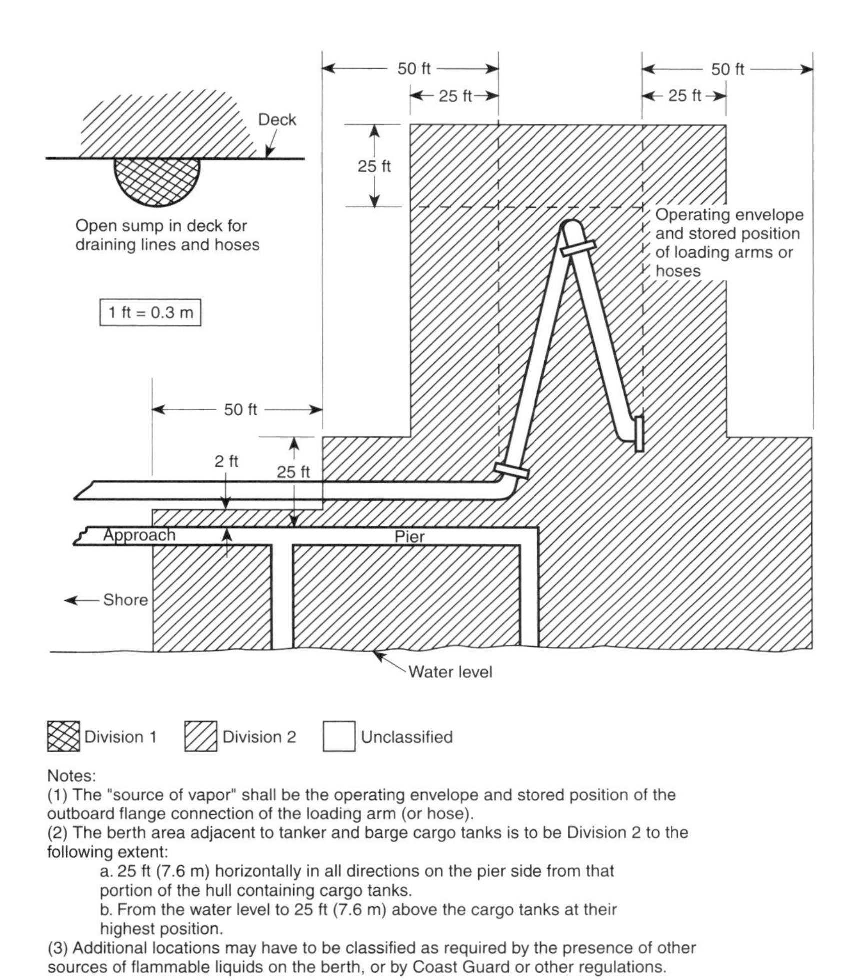

Notes:
(1) The "source of vapor" shall be the operating envelope and stored position of the outboard flange connection of the loading arm (or hose).
(2) The berth area adjacent to tanker and barge cargo tanks is to be Division 2 to the following extent:
a. 25 ft (7.6 m) horizontally in all directions on the pier side from that portion of the hull containing cargo tanks.
b. From the water level to 25 ft (7.6 m) above the cargo tanks at their highest position.
(3) Additional locations may have to be classified as required by the presence of other sources of flammable liquids on the berth, or by Coast Guard or other regulations.

Figure 3-27. *Classification of marine terminal at bulk storage plant.*

installations of storage batteries. It requires provisions to be made for sufficient diffusion and ventilation of gases to prevent an accumulation of an explosive mixture. It is relatively easy to comply with this requirement. Natural ventilation through upper wall or ceiling ventilation louvers is usually sufficient to dissipate any accumulations. Ventilation fans are not usually necessary.

Battery charging areas with adequate ventilation are not considered hazardous locations. Therefore, ordinary-location electrical equipment may be used. However, connections to battery terminals should not be made in a manner that would allow a spark to occur at the terminal. The amount of hydrogen produced is small, but its concentration would

Figure 3-28. Marine terminal handling flammable liquids.

be greatest at the battery vent. If the hydrogen concentration at the vent is in the explosive range, a spark near the vents could cause the battery to explode. Connections to switched devices should be made while the switch is in the off position.

If batteries used to power mobile equipment are to be charged, they should be connected to charging equipment in a manner that will not cause sparks at the terminals. Many lift trucks use extra-hard usage cables with insulated connectors, which can be attached to a mating connector on a cable attached to the charger. This prevents the connection from being made in the vicinity of the battery vents.

Automotive-type electric vehicles are becoming more common due to concerns about air pollution. Article 625 of the *National Electrical Code* covers charging equipment for automotive-type vehicles such as passenger cars, buses, trucks, vans, and the like.

If vented storage batteries are to be charged indoors, ventilation must be provided, as shown in Table 3-11. In addition, interlocks must be provided to prevent battery charging if the ventilation system is not operating properly. With unvented storage batteries, ventilation is not required.

3-2 Classification of Class II Hazardous Locations

3-2.1 General Guidelines

Class II locations are locations that are hazardous due to the presence of combustible dusts. Normally, such dust particles range in size from 0.1 to 1000 microns (1 micron = 10-6 m). These small particles burn very easily and rapidly, almost instantaneously. Within a building, a dust

Table 3-11 Minimum Mechanical Ventilation Required in Cubic Feet per Minute for Each Parking Space Equipped to Charge an Electric Vehicle

	Branch Circuit Voltage						
	Single-Phase			3-Phase			
Branch Circuit Ampere Rating	120 V	208 V	240 V or 120/ 240 V	208 V or 208 Y/ 120 V	240 V	480 V or 480 Y/ 277 V	600 V or 600 Y/ 347 V
15	37	64	74				
20	49	85	99	148	171	342	427
30	74	128	148	222	256	512	641
40	99	171	197	296	342	683	854
50	123	214	246	370	427	854	1066
60	148	256	296	444	512	1025	1281
100	246	427	493	740	854	1708	2135
150				1110	1281	2562	3203
200				1480	1708	3416	4270
250				1850	2135	4270	5338
300				2221	2562	5125	6406
350				2591	2989	5979	7473
400				2961	3416	6832	8541

explosion usually shakes other dust particles loose from the rafters and other parts of the structure, causing a secondary, usually more violent, explosion.

For the purpose of electrical classification, dusts fall into three general groups: E, F, and G.

3-2.1.1 Factors Affecting the Extent of Classification

Class II locations are not as easily classified as Class I locations are into Division 1 or Division 2. Gases and vapors tend to disperse as their distance from the source increases. The same may be true to a lesser extent with combustible dusts. Dust accumulations over time are greatest in the vicinity of equipment. However, these accumulations are obvious and are usually cleaned frequently. Dust will also accumulate on rafters and in concealed spaces of buildings. The rate of accumulation in these spaces is slower, but these spaces are cleaned less frequently. Dust collection can minimize the amount of dust released by a process, but housekeeping is still necessary. Ordinary ventilation may not be an effective means to minimize dust accumulation, as it may generate dust clouds. Strong ventilation may cause dust to accumulate elsewhere, such as on rafters, in equipment, below raised floors, and so forth.

Housekeeping should be based on the amount of dust that a process generates. Many dust-producing processes require a minimum of one cleaning per shift. Some processes require several cleanings per shift. Rafters and other concealed spaces should also be cleaned frequently. The most effective cleaning can be performed with soft brushes and vacuum cleaners suitable for Class II locations. Compressed air is often used to clean dust from concealed spaces and rafters of buildings. This is an extremely dangerous practice, as it generates the worst-case condition: a dust cloud and static electricity. Explosions have resulted from this type of housekeeping.

For conductive Group E, there are only Division 1 locations. The simplest means to minimize the hazard is to use only closed processing equipment and to contain the processing equipment in a closed room with self-closing doors. This may restrict the classified area to one room.

Nonconductive Group F and G dusts can be classified as Division 1 or Division 2. There are two criteria for determining the need for classification. The first criterion is whether or not there is a potential for ignition of a dust cloud. The mere presence of dust in the air does not necessarily imply that dust is present in that location in an explosive or ignitible mixture. (See Section 2-1.)

The second criterion is dust accumulation on surfaces. Dust that accumulates on surfaces tends to interfere with the proper dissipation of heat from equipment on which it accumulates. This may cause equipment to overheat. Dust accumulations may also interfere with the safe operation of equipment by penetrating joints and shafts.

Dust accumulations can become dust clouds if disturbed by a source of ventilation or vibration. NFPA 499, *Recommended Practice for the Classification of Combustible Dusts and of Hazardous (Classified) Locations for Electrical Installations in Chemical Process Areas*[23], uses depth of dust accumulation to delineate between Class II Division 1 and Division 2. An accumulation in excess of $^1/_8$ in. (3.18 mm) in a given area can be classified as a Class II, Division 1 location. If the accumulation is less than $^1/_8$ in. (3.18 mm), but there is some accumulation, the location can be classified as Class II, Division 2. If the accumulation is practically nonexistent, that is, the colors of surfaces are readily discernible, the location can be classified as nonhazardous.

3-2.2 Dust Collection Systems

Dust collectors are suction systems designed to collect powdered material. There are several types of dust collectors in common use, including cyclones, condensers, wet-type collectors, cloth-type and stocking arresters, and centrifugal collectors. The solid particles are collected in a hopper bin or silo. The material can be disposed of or recycled. Collectors may also be used to capture product. Collectors are usually located outdoors. Collectors used to collect very fine combustible dust and most metal dusts are usually wet-type collectors. Wet-type collectors pass dust particles through a water curtain, which causes the dust particles to form a sludge submerged in water. This sludge is not readily combustible.

The inside of a dust collector is normally classified as a Class II, Division 1 location. Wet-type collectors are normally considered nonhazardous downstream of the water curtain, unless they are used for combustible metal dust. The areas around dust collectors that require maintenance or cleaning can require classification as a Class II, Division 2 location if the maintenance allows dust to be released.

3-2.3 Static Electricity

A static charge can develop when two materials are separated. Charges can develop when dust is dispersed from a surface. The magnitude of the charge depends on a number of factors, including the size of the particle, the amount of surface contact, surface conductivity, gaseous breakdown, external field, and leakage resistance in a system. Static charge generation is greater when the dust is separated from a smooth surface rather than when the surface is rough. Charge generation is not likely to occur if both the dust and the surface on which it is resting are good electrical conductors. If the dust and the surface are of like materials, positive and negative charges are developed within the dust, resulting in a net charge of zero. When one material is metallic and the other is an insulator, the metallic material will usually develop a positive charge and the nonmetallic material will usually develop a negative charge. High humidity or grounding of the surface from which the dust is dispersed will not prevent electrostatic charge generation. Dust clouds and layers have been ignited experimentally by static discharge. In some experiments, the ignition has been caused simply by dust movement. In other experiments, the ignition has been caused by a static generator or other electronic equipment.

As with flammable liquids, there is a minimum concentration of dust that is necessary for ignition. Below this level, ignition cannot take place, regardless of the energy of the static discharge. The amount of energy required for ignition varies with the concentration of dust in the air. At higher dust concentrations, less energy is required to ignite the dust/air mixture. The minimum electrical energy to ignite a dust cloud is typically in the range of 10 to 80 millijoules. This low level of energy is less than the static that can be discharged from process machinery or from the human body. Layers of combustible dust can also be ignited, but no correlation has been found between ease of ignition of dust clouds and dust layers. Some metallic dusts require less energy for ignition than nonmetallic dusts.

Further information on the topic of static electricity can be found in Section 5-1.

3-2.4 General Equipment Requirements

The methods of equipment protection for Class II locations are different from those for a Class I location. A device that is approved for a Class I location is not necessarily approved for a Class II location. However, some devices have listings for both locations, because they have been tested and found to be suitable. The methods of equipment protection for Class II locations include dust-ignitionproof, dusttight, intrinsically safe, and pressurized.

Dust-ignitionproof equipment is enclosed in a manner that excludes ignitible amounts of dust or amounts that might affect performance or ratings. It is designed so that it will not permit arcs, sparks, or heat otherwise generated or liberated inside of the enclosure to cause ignition of exterior accumulations or atmospheric suspensions of a specified dust on or in the vicinity of the enclosure. For further information on dust-ignitionproof equipment, see Section 4-3.1.1.

Dusttight equipment is constructed so that dust will not enter the enclosing case under specified test conditions. For further information on dusttight equipment, see Section 4-3.1.1.

Intrinsically safe apparatus is equipment in which any spark or thermal effect, produced either normally or under specified fault conditions, is incapable of causing ignition of a mixture of flammable or combustible material with the air in its most easily ignitible concentration. Fur further information on intrinsically safe equipment, see Section 4-3.1.2.

Pressurized equipment is usually custom-built equipment. Requirements can be found in NFPA 496. For further information, see Section 4-3.1.3.

Article 502 of the *National Electrical Code* covers the requirements for electrical equipment in Class II locations. It is similar in format to Article 501 for Class I locations, in that requirements can be found in essentially the same locations in each article. However, the general requirements are not necessarily the same. A summary of some of the requirements follows. Further information on the various protection methods can be found in Chapter 4.

Switches, Circuit Breakers, and Motor Controllers: In Class II, Division 1 locations, switches are required to be provided with dust-ignitionproof enclosures. The switch and its enclosure must be approved as an assembly for Class II locations. If the switch is an isolating switch that is not intended to interrupt current during its normal operation, a tight metal enclosure, designed to minimize the entrance of dust, may be used in lieu of a dust-ignitionproof enclosure.

Tight metal enclosures must be equipped with telescoping or close-fitting covers or with other effective means to prevent the escape of sparks or burning material. In addition, the enclosure must have no openings (such as holes for attachment screws). This requirement is intended to prevent the escape of sparks or burning material that might ignite dust. In addition, it also prevents the entrance of dusts.

In locations where hazardous materials such as magnesium, aluminum, or aluminum bronze powders are present, the enclosures must be specifically approved for the type of dust.

In Class II, Division 2 locations, enclosures for switches, circuit breakers, and motor controllers must be dusttight. A dust-ignitionproof enclosure is unnecessary, but it is a permissible replacement.

Intrinsically safe circuits require very low energy levels. Therefore, intrinsically safe circuit breakers and motor controllers do not exist. However, a nondusttight switch for a limited energy control circuit can be installed in a Class II, Division 2 location if it is connected to an approved intrinsically safe device or connected to equipment in a safe location through a barrier with intrinsically safe outputs to the switch.

Transformers and Capacitors: Transformers and capacitors may represent a significant heat source. The insulating effect of a dust could cause overheating and consequent ignition of the dust layer. Installation of a transformer in a Class II location is heavily restricted. In a Class II, Division 1 location, the installation of transformers and capacitors containing liquid dielectrics that will burn is not permitted. A fire-resistive vault is equipped with self-closing doors on both sides of the fire wall that communicates with the Division 1 location. When a transformer does not contain a liquid that will burn, it may be installed in a vault, or it may be a unit that is approved as a complete assembly for Class II locations.

In Class II, Division 2 locations, transformers and capacitors containing combustible liquids must be installed in vaults. If a transformer contains askarel, it need not be installed in a vault. However, it must be provided with a pressure relief valve and a means to absorb generated gases or a means to vent these gases out of the building. In addition, a spacing of 6 in. to adjacent combustible materials must be maintained. The Environmental Protection Agency (EPA) has additional criteria that apply to askarel-filled transformers. These requirements severely restrict the use of any equipment containing askarel. The EPA regulations should be reviewed prior to any planned askarel-filled equipment installation.

Dry-type transformers rated 600 V or less, may be used in Class II, Division 2 locations if they are totally enclosed in a tight metal housing. The housing must not have ventilation openings, as this would permit dust to enter the enclosure. As an alternative, the dry-type transformer may be installed in a vault.

Transformer accessories containing arcing contacts, such as tap changes, should meet the requirements for switches in Class II locations.

Solenoids and Control Transformers: In Class II Division 1 locations, solenoids and control transformers, as well as associated switching mechanisms, must be installed in dust-ignitionproof enclosures. For a Class II, Division 2 location, solenoids and control transformers must be installed in tight metal housings without ventilation openings. The associated switching mechanisms must be installed in dusttight enclosures.

In a Class II, Division 2 location, a solenoid without switching or sliding contacts requires a tight metal enclosure to prevent dust accumulations on the coils.

Resistors: Resistors and resistance devices must be in dust-ignitionproof enclosures when installed in a Class II, Division 1 location. Dust-ignitionproof enclosures are also required for many Division 2 applications. However, in Division 2 locations, a tight metal housing may be provided in lieu of a dust-ignitionproof enclosure for nonadjustable resistors or resistors that are part of an automatically timed starting sequence, if the maximum operating temperature of the resistor does not exceed 120°C (248°F).

Motors and Generators: Motors and generators in Class II, Division 1 locations can be one of the following two types:

1. Dust-ignitionproof
2. Totally enclosed, pipe-ventilated

In order to use the totally enclosed, pipe-ventilated motor or generator design, the maximum surface temperature must not exceed the values in Section 500-3(f) of the *NEC*. In addition, the ventilation piping should be designed so that no dust will enter the piping.

Use of Flexible Cords: Flexible cord usage is slightly broader for Class II locations than it is for Class I locations. Flexible cord, which is approved for extra-hard usage, may be used on portable utilization equipment or on fixed equipment where flexibility is required. When flexible cord is used on fixed equipment in atmospheres containing conductive dust, it must be provided with dust-tight seals at both ends. For portable equipment, the seal must be provided at boxes or fittings required to be dust-ignitionproof. These requirements apply regardless of whether the location is Division 1 or Division 2.

Locations Requiring Classification as Both a Class I and a Class II Location: Occasionally, a location exists where both Class I and Class II hazards are present. Plants with dust-producing processes often need to use materials such as flammable liquids or gases. There are also some dusts, such as coal dust and grain dust, that liberate methane gas.

Equipment suitable for Class I only or Class II only is obviously not suitable, since the protection methods are designed to prevent a very specific ignition scenario. However, equipment that is suitable for both environments may also be unsuitable if materials with low autoignition temperatures are in use. Equipment that is layered with dust may produce higher temperatures than it would in the unlayered condition. Such a condition could limit the use of the device in the Class I environment. In order for a device to be acceptable, the maximum surface temperature must not be capable of igniting either atmosphere.

3-2.5 Classification of Specific Processes/Occupancies

Standards, guidelines, and recommended practices exist for the classification of Class I hazardous locations. Many of these documents provide usable guidelines on the limits of the Division 1 and the Division 2 areas. When an area is hazardous due to the presence of a flammable gas or vapor, the air-gas mixture becomes lean as the distance from the source increases. Although there may be a strong odor in the air, the distance from the source of the air-gas mixture may be so great that the mixture is not combustible. A good source of ventilation may dilute the air-gas mixture to the point that the hazardous area is considerably reduced.

The classification of Class II locations can be considerably more difficult. Dusts do not readily disperse in the same way that gases or vapors tend to. Dusts accumulate and are heaviest in the immediate vicinity of processing equipment. The accumulations taper off as the distance from the processing equipment increases. As indicated earlier in this chapter, a ventilation system can create a hazard if it causes dusts to be dispersed in a cloud. Dust-collection systems can be used to minimize accumulation of dust, but some other means of housekeeping is usually necessary. Dust accumulations in the immediate vicinity of processing equipment are often evident and usually are removed. However, dust usually accumulates on structural steel, behind walls and partitions, and in and on other machinery.

The other important consideration for dusts is their insulating effect on equipment. A layer of dust will reduce the ability of a device to radiate heat. This may cause the device to develop excessive temperatures. Some dusts will carbonize if they are in contact with hot surfaces for an extended period of time. A carbonized dust will have a reduced ignition temperature.

The classifications discussed in the following paragraphs are based on NFPA 499 and existing standards and industry practices. Where existing standards do not provide a classification, the authors suggest criteria that warrant classification.

3-2.5.1 Facilities Manufacturing and Handling Flour or Starch

The processing and handling of flour or starch can be very dangerous. Its fine, powdery form makes it easily ignitible. It also becomes airborne easily. Agricultural commodities, such as soybeans and wheat, corn, and other grains, are the typical raw materials that are processed into flour or starch.

Starch is a granular or powdery complex carbohydrate. It is usually made from corn, but it may also be a combination of other materials and produced as a low-grade flour. It is typically odorless and tasteless. In various forms it may be used as a food additive, in adhesives, or as a pharmaceutical additive. It may also be used in textile processes as a sizing and finishing material and as a thickener for textile printing colors.

Flour is a finely ground meal, usually of wheat. It may consist of other additives such as barley and corn. It is available in various grades. The higher grades are usually used for baking products, such as bread.

These basic raw materials are ground or pulverized into fine powder in milling machines. The type of machinery used depends upon the degree of fineness needed. Roller mills are typically used to grind wheat and rye into high-grade flour. A roller mill pulverizes the feed material between two or more rollers revolving at different speeds. The ground material is removed from the last roller by a scraper blade.

High-speed hammer mills or pin mills are used to produce flour with a controlled protein content. Hammers grind the feed against a grinding plate. Pin mills consist of two disks on which a series of pins are mounted in a circular arrangement. One or both of the disks may rotate. The commodity is ground between the pins.

Grinders are supplied with feed and then discharge ground product via conveying systems. Most of these conveying systems are pneumatic. The grinding equipment and the conveyors are usually closed systems. The closed system is intended to minimize the release of dust into the environment. However, in normal operation, some dust will be released. Areas subject to a limited release of dust may warrant classification as a Class II, Division 2 location with a good housekeeping program. If dust is allowed to accumulate, a Division 1 classification may be more appropriate.

The raw material for the product typically arrives at the mill in a railroad boxcar or a hopper car. Wheat and other commodities are usually removed from the car through a pneumatic conveying system. The pneumatic conveying system conveys the product through a stream of high-velocity air from the boxcar (or hopper car) to an elevator, and from there into a storage silo. A static charge can be generated in the commodity during the transfer. If the charge becomes too great, the cloud of dust in the duct could be ignited. Therefore, NFPA 61, *Standard for the Prevention of Fires and Dust Explosions in Agricul-*

tural and Food Products Facilities[24], requires the pneumatic tubing to be electrically conductive. The potential between the boxcar and the plant is further stabilized by bonding the boxcar to the plant's system. (See Figures 3-29, 3-30, and 3-31.)

The pneumatic tubing is a closed system to prevent the escape of product. Air is recirculated through the system and the surrounding environment through filters in an air-material separator. Some fugitive dust may escape in the process, but most of it will be collected by dust collection systems. Areas in which fugitive dust accumulates may warrant classification as a Class II, Division 1 location.

The point at which the air-fuel mixture of combustible dusts becomes too rich has not been determined experimentally. Therefore, the inside of a conveying tube or duct is usually classified as Class II, Division 1.

Other product-conveying systems may be used within the plant.

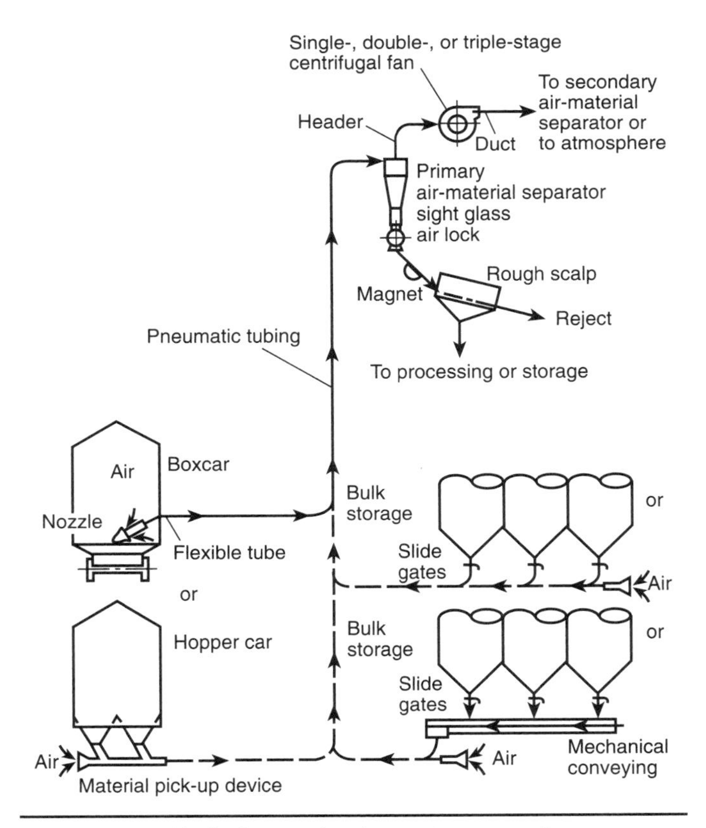

Figure 3-29. *Typical car unloader system, negative pressure-type, low capacity.*

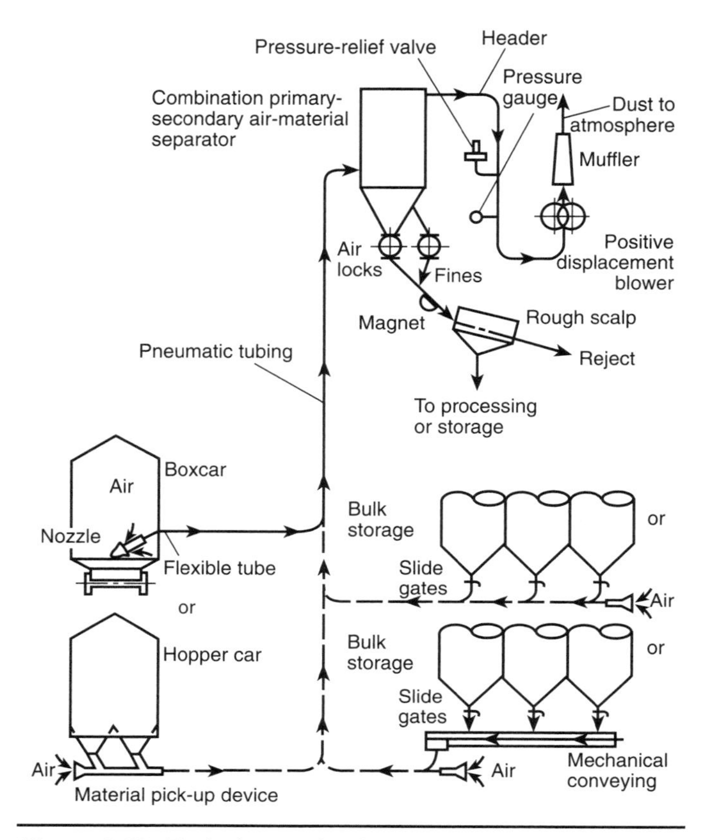

Figure 3-30. *Typical car unloader system, negative pressure-type, high capacity.*

Common systems include bucket elevators, screw conveyors, and belt conveyors. Bucket elevators are usually used to raise a product vertically for discharge into a silo. They are the only practical method to transport feed into the taller silos. Bucket elevators are inherently a very dusty means of product transportation because product tends to leak out of the buckets. In addition, product intended to be discharged from buckets at the top of the elevator may form dust clouds as it falls to the bottom of the silo. Therefore, the area in the immediate vicinity is usually classified as a Class II, Division 1 location.

Screw conveyors and conveyor belts are usually used to transport a product horizontally. Screw conveyors are often used to feed product into machinery. Although screw and belt conveyors do not produce as much dust as pneumatic systems, the ductwork through which they transport product is often classified as a Class II, Division 1 location.

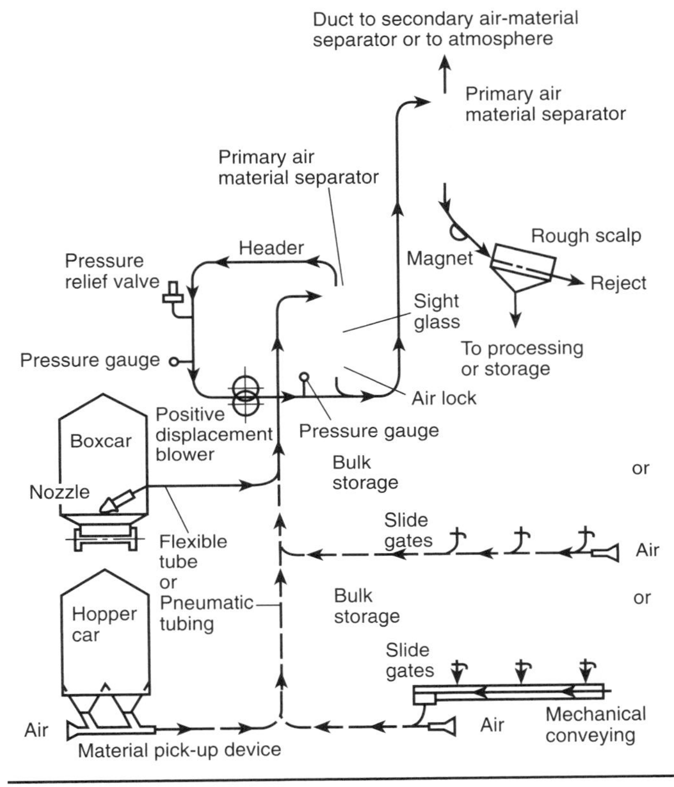

Figure 3-31. *Portable car unloader and transfer system, combination negative pressure-type and positive pressure-type, high capacity.*

In this instance, the classification is due to layers of product that can accumulate on any equipment installed therein. The need to install equipment in these spaces is quite limited and usually consists mostly of transducers, thermocouples, and the like. Any other equipment is usually located outside of the ductwork.

Tramp metal in a conveying system can be a serious hazard if it passes into equipment where it could cause mechanical damage, such as grinding equipment. It may cause sparks, which could ignite the dust. Tramp metal detectors and magnetic separators are installed to prevent the passage of tramp metal into machinery.

Many feed products are abrasive to the ductwork through which they are transported. This is particularly true of pneumatic systems where the product continuously impacts the tubing. This abrasion problem may require frequent maintenance of ductwork or tubing to prevent a large-scale release of product into the plant. Spouts used to channel the flow of grain into elevators are usually fitted with abrasion-resistant liners. When a liner is damaged, it can be replaced without requiring the replacement of the entire spout. The areas where potential leaks from ducts may occur may warrant classification as Class II, Division 2 locations. If trace amounts of dust exist at that location, even when they are frequently removed through regularly scheduled housekeeping, the area may warrant classification as a Class II, Division 2 location. If housekeeping becomes infrequent, the area may warrant a change in classification from a Division 2 location to a Division 1 location.

Grain often has a high moisture content when received at the elevator. It usually requires processing through a dryer to preserve the quality of the grain. However, drying the grain tends to make it more combustible.

Grain must also be screened, cleaned, or scalped to remove impurities such as seeds, pods, and husks. These materials are usually drier and more ignition-prone than the grain itself. An air-aspiration system is used to remove the dust generated by these unwanted materials. The dust generated is usually removed through a dust collector located outside of the building.

Other agricultural commodity processing facilities, such as food plants, may pose similar dust-generation problems. The problems are usually more pronounced in plants using batch-type processes than they are in continuous-process plants. Continuous processes are usually closed systems. Batch-process plants often involve processing steps in which commodities are added to mixers from bags, drums, or hoppers. This may result in considerable dust generation. The area in and around these mixers usually requires a great deal of housekeeping. Motors and other electric equipment associated with mixers may need to be suitable for Class II, Division 1 or perhaps only for Class II, Division 2 locations. The actual classification depends on the degree of exposure to the dust hazard.

Examples of food-processing plants with similar hazards include sugar plants, cake mix plants, gelatin plants, bakeries, and so forth. Many of these facilities involve dust-producing commodities and dust-producing processes. The most important variable in the area classification scheme for one of these occupancies is the highest degree of housekeeping. Dust accumulations are usually greatest at the processing equipment itself. Dust particles tend to drift from the primary processing area to areas remote from the process. The prevailing ventilation in the plant may cause substantial accumulations in other areas, including structural beams and concealed spaces. These other plant areas also require frequent housekeeping, because accumulations in these areas may eventually interfere with the safe operation of equipment. These accumulations may also provide fuel for secondary explosions.

3-2.5.2 Coal-Handling and Storage Facilities

Coal is a carbonaceous material that can be explosive in powdered form. The hazard exists in mines and processing facilities.

There are two principal groups of mines: underground mines and surface mines. The area classification and electrical installation requirements for underground mines are outside the scope of the *National Electrical Code* and will, therefore, not be covered here. These facilities are governed by the Mine Safety and Health Administration (MSHA). Surface mines and other facilities that handle coal products are usually governed by the requirements of the *National Electrical Code*.

Coal may present a dust-explosion hazard and, in some instances, a gas explosion hazard. It may also begin burning through a chemical reaction known as spontaneous heating. Coal is mined primarily for use as a fuel. For most industrial applications, the coal is pulverized into fine particles, which can be mixed with air and sprayed into a burner for ignition.

Coal is generally stored in outside storage piles, bunkers, silos, and bins. Freshly mined or recently crushed coal tends to liberate quantities of methane and some other gases. When stored in a bunker or silo, the bottom of the housing is usually sealed off to prevent entrance of air. This is to prevent oxidation, which can lead to spontaneous heating of the coal. Ventilation is usually provided for the space above the coal to allow the gases to dissipate.

Classification of coal handling and storage areas depends on the storage conditions and the type of occupancy. In coal preparation plants, NFPA 120, *Standard for Coal Preparation Plants*[25], provides requirements for area classification. These requirements are summarized in Table 3-12. The requirements apply only to coal preparation plants. In other areas where coal is stored or used, this table cannot be used. In these areas, some judgments of the prevailing conditions must be made.

In outdoor storage pile areas and in plant areas of open construction, natural ventilation may be sufficient to prevent the accumulation of gases or dust. These areas are normally classified as nonhazardous. However, in bunkers, bins, and silos containing freshly mixed or freshly pulverized coal, some classification may be necessary. If adequate ventilation is provided, the area can be classified as a Class I, Division 2 location. If failure of the ventilation system would not result in an accumulation of an explosive or ignitible mixture of methane or any other gas, the area can be considered nonhazardous with respect to the Class I hazard. If adequate ventilation is not provided and explosive or ignitible mixtures exist intermittently or continuously, the location would be classified as a Class I, Division 1 location. However, where these conditions should not be allowed to continue to exist, ventilation should be provided.

As previously indicated, the dust hazard must also be evaluated.

Table 3-12 Electrical Classification of Coal Preparation Plants

Location	Classification
Plant areas of open construction where coal dust or any combustible gases are freely dispersed to open atmosphere	Nonhazardous
Control rooms, electrical equipment rooms, and substations that are provided with adequate ventilation to prevent combustible gas or coal dust accumulations	Nonhazardous
Enclosed areas where coal dust is kept sufficiently wet to prevent dust from becoming airborne	Nonhazardous
Enclosed areas where failure of ventilation would result in the accumulation of methane gas in ignitible quantities	Class I, Division 2, Group D
Enclosed areas where coal dust will not normally be in suspension in the air or in explosive or ignitible quantities or where coal dust may be in suspension due to equipment malfunction	Class II, Division 2, Group F
Areas in which ignitible concentrations of dust may be in suspension in the air in explosive or ignitible quantities	*Class II, Division 1, Group F
Areas in which ignitible concentrations of methane may be present in the air in explosive or ignitible mixtures.	*Class I, Division 1, Group D

**May require classification as a Class I and Class II location. The National Electrical Code requires electrical equipment that has been evaluated for simultaneous exposure to both atmospheres.*

When the coal supply consists mostly of large nuggets of hard coal, dust is present in small amounts. If the dust accumulations in the area of equipment are minimal, the area can warrant a classification of Class II, Division 2, Group F. If there is no discernible dust, the area may be classified as nonhazardous. An accumulation of dust of less than $^1/_8$ in. (3.18 mm) is usually considered to be a Class II, Division 2 location. When the amount is $^1/_8$ in. (3.18 mm) or greater, the location is classified as Class II, Division 1.

Nuggets are often unsuitable to feed directly into an industrial boiler or furnace. The coal nuggets must be pulverized into a finer form. The pulverized coal can then be mixed with air and fed directly from the pulverizer into the burner. It can also be ground in a grinding system and stored in bins or hoppers until it is needed.

In locations where completely dusttight pulverizing systems are used, the locations may warrant classification as nonhazardous. However, many pulverizing systems emit some coal dust. The actual classification of the area is a function of how much coal dust may be present.

The areas where the greatest dust accumulations or dust clouds may be present include those around conveyors. Where dust clouds are not present and not likely to be present, the classification may be determined by measuring the depth of dust. Trace amounts of dust may be classified as Class II, Division 2 locations. Where the depth of dust is $^1/_8$ in. (3.18 mm) or greater, the location should be classified as Division 1.

Where the pulverized coal bin storage system is used, coal may be stored in closed bins. If bins are removed from the storage area to supply the furnace, the storage area may be classified as a Class II, Division 2, Group F location. Where coal is removed from bins using a process that generates dust, (e.g., shoveling from bins, dumping of hoppers, and so on), the area around the dumping operation might warrant classification as a Class II, Division 1, Group F location. The extent of the classification depends on how much dust will be generated. Frequent housekeeping, such as a dust-collection system, could mitigate the extent of the hazard due to accumulations.

A location that might otherwise warrant classification as a Class II, Division 1 or Division 2 location would be classified as a Class II, Division 1 location if the coal dust in question is electrically conductive. This is because electrical equipment suitable for a nonconductive dust may not be suitable for a conductive dust. An electric current tends to find the path of least resistance through a conductive dust layer, heating up the dust particles in its path and thus providing a source of ignition. An arc between terminals through the dust may occur, igniting a dust cloud or a dust layer. This phenomenon is commonly known as tracking.

3-2.5.3 Processes That Produce Combustible Conductive Dust

Electrically conductive dusts are particularly hazardous. Metallic dusts such as aluminum powder, magnesium powder, and atomized aluminum produce the most violent explosions among the combustible dusts. Electrical equipment suitable for a nonconductive dust environment may not be suitable for a conductive dust environment. Although dusts such as aluminum rapidly form oxides, the oxides can be broken down, rendering the dust conductive. Higher voltages break down the oxides on metal dusts faster. As the oxides begin to break down, a current flows through the dust, causing the dust to heat. This may cause ignition of the dust layer. This phenomenon is known as tracking. In addition, when a conductive dust exists in a cloud, it may be readily ignited by arcing. When combustible metal dusts are released into the atmosphere or layered on equipment, it is assumed that simultaneous failure of ordinary equipment could occur. Therefore, it is assumed that only Division 1 locations can exist.

Combustible metal dusts exist due to a number of different processes. The dust produced may be a by-product of a process, such as drilling, grinding, milling, polishing, or buffing. The dust also may be manufactured for a powdered metallurgy product, such as aluminum or magnesium powders. However, in machine shop operations such as drilling or lathing clouds of metal dust are very rare. Most of the particles released are chips and fines. These particles are usually too heavy to be conveyed a great distance from the machinery. Some smaller particles will be generated, but since most of these operations are conducted with cutting oils, the smaller particles will not be released as dust into the atmosphere. Grinding and polishing release smaller dust-type particles than do machine shop processes. These operations can become hazardous if the dust is allowed to accumulate in hazardous quantities.

The greatest dust hazards are associated with powder metallurgical processes. Powder metallurgy is the manufacture of products from finely ground metal particles. The two principal processes include powders as additives and powder pressing processes. In the powder additive process, metal powders are blended with other materials to change the characteristics of the final material. For example, powders of aluminum and bronze are added to paints to provide a metallic finish. Metallic powders may also be added to fire-retardant materials. Powder additives include metal powders used in the plastics industries. They may be added to plastics to increase the strength or wear resistance of the plastic. They also may be added to increase the conductivity of the plastic.

One of the largest applications of powder metallurgy is the manufacture of parts made from high-temperature refractory metals such as tungsten, molybdenum, and tantalum. These metals melt at such high temperatures that it is very difficult to melt appreciable quantities of these materials into molded parts. For these applications, the refractory material is ground into fine particles. The particles can then be injected into a die or mold, or they can be slip cast or formed into shapes by some other means. Metal powder can be cast dry or dispersed in a liquid. Some powders are mixed with thermoplastic materials, which may be removed during sintering. The powder is then formed into the desired shape by pressing it into a die or mold at high pressure. It may utilize a centrifugal process, or it may use a process such as a hot isostatic press. Many powder-cast products are heat treated for higher strength. It is often necessary to perform some final grinding, sanding, or polishing to achieve the necessary tolerances or finish on the end product. These processes are common in several industries. Tools requiring hardness and high-strength are often made from powder metallurgy. Large, high-strength parts, such as aircraft landing gear struts, are made from powdered titanium.

The area classification scheme for metal-dust-producing processes requires careful evaluation. Sections 500-6(a) and 502-1 of the *National Electrical Code* indicate that, where dusts of an electrically conductive nature are present, there are only Division 1 locations. In addition, the fine-print note (FPN) following Section 500-6(a) calls special attention to the hazards of magnesium and aluminum dusts.

A classification scheme should follow the normal guidelines for Class II, Division 1 and Division 2 locations. However, the areas that are normally considered Division 2 locations are considered Division 1 locations. The classification scheme does not need to go to the extreme of classifying nonhazardous locations as Division 1 locations.

Areas where drums, sacks, bins, or other containers of powder are stored may warrant classification as Class II, Division 1, Group E locations. This recommendation is based on the assumption that the handling of these containers can result in the release of quantities of dust. If the container is dropped or damaged, a cloud of dust could be released.

Some areas may not warrant classification as Class II locations. These include areas around drilling machinery, lathes, and similar equipment where most of the particles released are turnings, chips, and larger chunks of metal. A notable exception is magnesium processing, in which chips, turnings, and similar particles are produced. Even though these particle sizes are too large to be considered dusts, the particles are very flammable.

Areas that may warrant classification as Class II, Division 1, Group E locations include those around milling machines that reduce metals into fine powders and others with a similar function. The extent of the classification depends on factors such as the effectiveness of dust collection equipment, good housekeeping, and the presence of interlocks. An effective dust-collection system combined with good housekeeping should prevent dust from being released in significant quantities into the area, as well as prevent any accumulations.

The inside of any equipment that mixes or blends combustible metal powders is considered a Class II, Division 1 location. The extent of the area classification around batch processes may be larger than around continuous processes. Batch processes often permit more dust to be released, because the product is often transferred from drums. Continuous processes usually transfer product via closed conveyors, ducts, or piping systems.

In order to minimize the extent of area classification, metal powder processes are usually kept in small, isolated rooms. The rooms are usually equipped with self-closing doors, which are kept closed. The best area for such a room is near an exterior wall, because dust-collection system ducts passing through other areas of the building may cause dust to enter other areas.

3-3 Classification of Class III Locations

3-3.1 General Guidelines

Class III locations are locations that are hazardous due to the presence of easily ignitible fibers or flyings but in which such fibers or flyings are not likely to be in suspension in the air in quantities sufficient to produce ignitible mixtures. These locations include the following:

1. Class III, Division 1 locations where easily ignitible fibers or materials producing combustible flyings are handled, manufactured, or used
2. Class III, Division 2 locations where easily ignitible fibers are handled or stored, excluding handling that occurs during the manufacturing process

3-3.1.1 Factors Affecting the Extent of a Class III Location

In Class III locations, there are no material groupings. Material groupings in Classes I and II are used to separate materials with ignition characteristics that are easily affected by the construction of electrical equipment. There is no such condition in Class III locations. The fibers are too large to penetrate flanged joints and are not electrically conductive.

Class III locations are further distinguished from Class I and II locations in that the materials do not represent an explosion hazard. These materials are easily ignitible and will burn rapidly.

Typical Class III locations include some parts of cotton, rayon, and other textile mills, combustible fiber manufacturing and processing plants, cotton gins and cottonseed mills, clothing manufacturing plants, woodworking plants, and so forth.

3-3.2 Equipment — General Requirements

Requirements for Class III locations can be found in Article 503 of the *National Electrical Code*. Most of the requirements are the same for Division 1 and 2 locations. The only exception is the permissible wiring methods. Open wiring on insulators is permitted in Class III, Division 2 locations, provided that the section, compartment, or area contains no machinery and conductors are run in roof spaces or protected in accordance with Section 320-14 of the *NEC*.

Boxes and fittings for Class III locations are required to be dusttight. This requirement was new with the 1987 edition of the *NEC*. Prior editions required telescoping or close-fitting covers or other means to prevent the escape of sparks or other burning material. This made equipment specification difficult. Dusttight equipment is readily available and has the necessary characteristics to prevent the escape of sparks or burning material. Transformers, switches, circuit breakers, motor controllers, control transformers, and resistors are all required to be installed in dusttight enclosures. There are no requirements for conduit seals between dusttight enclosures in Class III locations.

The requirements for motors and generators in Class III locations are similar to those for Class II, Division 2. The motor or generator is required to be totally enclosed nonventilated, totally enclosed pipe ventilated, or totally enclosed fan cooled.

In locations where the authority having jurisdiction judges that only moderate accumulations of lint or flyings will occur, and where the machinery is readily accessible for cleaning, the following are permitted:

1. Self-cleaning, squirrel-cage textile motors
2. Standard open-type machines without sliding contacts, centrifugal, or other types of switching mechanisms, including motor overload devices
3. Standard open-type machines having switching contacts, switching mechanisms, or resistance devices enclosed within tight housings without ventilation or other openings

Lighting fixtures within Class III locations must be of a type designed to minimize the entrance of combustible fibers and flyings. They must also be designed to prevent escape of sparks, burning materials, or hot metal. Guarding is required for fixtures that are subject to physical damage. Complete requirements can be found in the *National Electrical Code*.

3-3.3 Textile Facilities

Textile facilities can house a number of operations that could be classified as Class III locations. Class III operations are characterized by the presence of lint and similar flying material. There are no NFPA standards on textile manufacturing facilities. Therefore, the classification of the areas around each of the operations involves careful evaluation of the likelihood that a given process or operation will produce combustible flyings. As in Class II locations, a good housekeeping program will help to limit the extent of the hazardous location.

Some operations present obvious hazards. Opening, picking, and blending are operations in which loose fibers move rapidly through machinery. Some fibers tend to separate and become airborne. Some other operations, such as storage of raw material, are less obvious but are still hazardous.

Cotton and similar materials are shipped to the textile plant in bales. The bales are usually stored in a storage area prior to processing. Each time bales are handled in this area, some fibers may separate. This area may warrant classification as a Class III, Division 2 location, because it meets the definition of Section 500-7(b) of the *NEC*. Without proper housekeeping, the area could become a Division 1 location.

The first manufacturing operation is the insertion of bales into opening equipment. This equipment opens bales and separates foreign matter from the desired product. The classification of the area around the equipment depends on whether the equipment is fully enclosed. At the loading end, some fibers will be liberated from the bales. Generally, the classification of this area is Class III, Division 1.

After bales are opened, fibers are conveyed to the picking and blending feeder. A picker normally consists of a condenser, rake beater sections, and lap rolls. In this equipment, fibers are cleaned and compressed into sheets. In modern mills, this equipment is enclosed and equipped with dust collectors. If the machinery is not totally enclosed, some fibers may be released. The areas where fibers are released may warrant classification as Class III, Division 1 locations.

Weaving is a process in which fibers are interwoven on equipment, such as looms. Fibers will generally be released during weaving. Therefore, this area may merit classification as a Class III, Division 1 location.

Rolled cloth is usually passed through large dyeing or printing machines to add color or patterns. Normally, high-flash-point inks are used to eliminate the need for classification as a Class I location. If low-flash-point inks or dyes are used, then the area should be evaluated as a Class I location. The dyeing and printing machinery may consist of several operations, including washing and drying of the cloth. These operations are intended to minimize material shrinkage. Some parts of the machinery may tend to produce lint, while other operations, such as washing, will not. The area in and around the equipment should be carefully evaluated to determine which parts of the equipment are likely to release lint. The areas where lint is released may warrant classification as Class III, Division 1 locations. Dust collection systems should be designed to remove as much of the lint as practical.

Textile weaving and dyeing plants normally produce products in the form of rolled sheets. The rolls are sold to manufacturers, who then cut the cloth into patterns. The cut cloth is sewn together into finished products, such as clothing. The cutting and sewing operations generate a considerable amount of lint. These areas normally warrant classification as Class III, Division 1 locations.

The area in which rolled textiles are stored will normally be classified as Class III, Division 2 locations.

Bibliography

[1]NFPA 70, *National Electrical Code®*, National Fire Protection Association, Quincy, MA, 1996.

[2]NFPA 30, *Flammable and Combustible Liquids Code,* National Fire Protection Association, Quincy, MA, 1996.

[3]NFPA 496, *Standard for Purged and Pressurized Enclosures for Electrical Equipment*, National Fire Protection Association, Quincy, MA, 1993.

[4]NFPA 497, *Recommended Practice for the Classification of Flammable Liquids, Gases, or Vapors and of Hazardous (Classified) Locations for Electrical Installations in Chemical Process Areas,* National Fire Protection Association, Quincy, MA, 1997.

[5]ANSI/UL 87, *Power Operated Dispensing Devices for Petroleum Products*, Underwriters Laboratory, Northbrook, IL, 1990.

[6]NFPA 30A, *Automotive and Marine Service Station Code*, National Fire Protection Association, Quincy, MA, 1996.

[7]*National Electrical Code Handbook,* National Fire Protection Association, Quincy, MA, 1996.

[8]NFPA 88B, *Standard for Repair Garages*, National Fire Protection Association, Quincy, MA, 1997.

[9]NFPA 302, *Fire Protection Standard for Pleasure and Commercial Motor Craft*, National Fire Protection Association, Quincy, MA, 1994.

[10]NFPA 88A, *Standard for Parking Structures*, National Fire Protection Association, Quincy, MA, 1995.

[11]NFPA 407, *Standard for Aircraft Fuel Servicing*, National Fire Protection Association, Quincy, MA, 1996.

[12]NFPA 33, *Standard for Spray Application Using Flammable or Combustible Materials*, National Fire Protection Association, Quincy, MA, 1995.

[13]NFPA 34, *Standard for Dipping and Coating Processes Using Flammable or Combustible Liquids*, National Fire Protection Association, Quincy, MA, 1995.

[14]NFPA 86, *Standard for Ovens and Furnaces,* National Fire Protection Association, Quincy, MA, 1995.

[15]NFPA 86C, *Standard for Industrial Furnaces Using a Special Processing Atmosphere,* National Fire Protection Association, Quincy, MA, 1995.

[16]NFPA 45, *Standard on Fire Protection for Laboratories Using Chemicals,* National Fire Protection Association, Quincy, MA, 1996.

[17]NFPA 99, *Standard for Health Care Facilities*, National Fire Protection Association, Quincy, MA, 1996.

[18]NFPA 58, *Standard for the Storage and Handling of Liquefied Petroleum Gases*, National Fire Protection Association, Quincy, MA, 1995.

[19]AGA *Recommended Practice for Gas Compressor Stations*, XFO 277, American Gas Association.

[20]NFPA 59, *Standard for the Storage and Handling of Liquefied Petroleum Gases at Utility Gas Plants*, National Fire Protection Association, Quincy, MA, 1995.

[21]NFPA 59A, *Standard for the Production, Storage, and Handling of Liquefied Natural Gas (LNG)*, National Fire Protection Association, Quincy, MA, 1996.

[22]NFPA 32, *Standard for Drycleaning Plants*, National Fire Protection Association, Quincy, MA, 1996.

[23]NFPA 499, *Recommended Practice for the Classification of Combustible Dusts and of Hazardous (Classified) Locations for Electrical Installations in Chemical Process Areas,* National Fire Protection Association, Quincy, MA, 1997.

[24]NFPA 61, *Standard for the Prevention of Fire and Dust Explosions in Agricultural and Food Products Facilities*, National Fire Protection Association, Quincy, MA, 1995.

[25]NFPA 120, *Standard for Coal Preparation Plants*, National Fire Protection Association, Quincy, MA, 1994.

CHAPTER 4

Equipment Protection Systems

4-1 General Concepts

When electrical equipment is installed in a hazardous location, precautions are necessary to prevent the equipment from igniting the flammable atmosphere, propagating fire or explosion.

All equipment protection systems are designed to prevent ignition of the flammable atmosphere by removing one or more legs of the fire triangle. (See Section 2-8 and Figure 2-7.) Protection systems do this in different ways, as noted below. Each of these systems or techniques will be explained in this chapter.

Purged and pressurized systems separate the fuel from the ignition source inside the purged and pressurized enclosure. They may also remove the air or oxygen.

Intrinsically safe systems prevent ignition by maintaining the available energy below the level needed to cause ignition; that is, they remove the ignition source.

Dust-ignitionproof systems prevent ignition by separating the fuel (dust) from the ignition source.

Explosionproof and flameproof systems prevent ignition by isolating the ignition source from the flammable atmosphere outside the explosionproof or flameproof enclosure, while at the same time anticipating that the flammable atmosphere will be ignited inside the enclosure.

Increased safety systems prevent ignition by preventing ignition sources (sparks, arcs, or high temperatures); that is, they remove the ignition source.

Oil-immersed, hermetically sealed, powder- (sand-) filled, and encapsulated systems prevent the ignition source from contacting the flammable atmosphere.

4-1.1 Class I Locations

4-1.1.1 Division 1

Class I, Division 1 hazardous locations are those locations where flammable gases or vapors are likely to be present under normal conditions, either continuously or intermittently. (See the definition in Section 2-4.3.2.1) The design of the protection systems for Division 1 locations therefore is based on the flammable atmosphere being present continuously, as this is the worst-case condition.

Groups: Explosionproof and intrinsically safe protection systems, both of which are suitable for protection in Class I, Division 1 locations, are sensitive to the combustion properties of the particular atmosphere present. Therefore, they must be designed for the particular Class I group or groups of materials in which they are intended to be used. Purged and pressurized, hermetically sealed, and oil-immersed protection systems are insensitive to the particular material involved, since they separate the fuel from the ignition source and are thus suitable for all Class I groups (at least in theory). However, even though the particular protection system may not be sensitive to the combustion characteristics of the flammable material, the entire protection system may be a combination of systems. For example, a purged and pressurized enclosure requires a pressure- or flow-sensing device that will detect a loss of the protecting flow or pressure and disconnect the internal electrical parts and/or, in some applications, initiate an alarm. This sensing system is often electrical and may itself be protected by explosionproof enclosures or intrinsically safe circuits, thus making the entire protection system suitable for only those groups of Class I hazardous locations for which the sensing system is suitable.

Temperature: Temperature is another consideration. Some types of equipment operate at relatively high external surface temperatures under normal operating conditions, and the equipment, if it operates at over 212°F (100°C), must be marked to indicate its maximum operating temperature. Other types of equipment operate at elevated but not ignition-capable temperatures under normal conditions, but they may reach much higher temperatures under abnormal but not necessarily unusual conditions. Motors are one example, as they can be, and often are, overloaded. In addition, bearings can fail and other faults can result when three-phase motors are single-phased. Transformers also can be overloaded, and solenoid valves can stick in the unenergized position. Other types of equipment usually are less likely to have high external surface temperatures, because a fault condition, such as an arcing fault to the enclosure, is necessary before temperatures outside of the protecting enclosure can reach or approach the ignition temperature of the surrounding flammable atmosphere. Junction boxes and other parts of the wiring system, such as raceways, are examples of this type of equipment. When investigating the protection systems, qualified electrical testing and approval agencies take all of these factors into consideration.

4-1.1.2 Division 2

Because flammable gases and vapors are not present under normal conditions in Division 2 locations, and they are likely to be present only for relatively short periods under abnormal conditions, protection system requirements for Class I, Division 2 locations need not be as stringent as they are for Class I, Division 1 locations. The general concept in Class I, Division 2 locations is that if the electrical equipment constitutes an ignition source under normal electrical equipment operating conditions, the protection system essentially should be the equivalent of a protection system in a Division 1 location, taking into account the likelihood that the flammable mixture will not be present for an extended period of time. For example, a switch in a 480-V power circuit is an ignition source under normal operating conditions of the switch because of the arc at the switch contacts when it opens or closes the circuit. These arcing contacts, therefore, require special attention, even in Division 2 locations. On the other hand, a splice in a junction box is not an ignition source under normal conditions for the splice, and special precautions are not needed for splice enclosures in Class I, Division 2 locations.

If a special protection system is necessary because the electrical equipment or part of the electrical equipment is an ignition source under normal conditions, this protection system can be designed so that it will be suitable over a relatively long period of time, but not necessarily forever. In Division 2 locations, the flammable atmosphere is not likely to be present for very long. Thus, the switch contacts may be enclosed in a hermetically sealed enclosure in Division 2 locations, but not in Division 1 locations. A hermetically sealed enclosure is, in effect, an enclosure with an extremely slow leak rate. However, for most hermetically sealed enclosures, there is no way of knowing at any moment in time whether the hermetic seal is still effective. If a hermetically sealed enclosure were permitted to enclose arcing contacts in a Division 1 location, where flammable atmospheres are anticipated continuously, the atmosphere could eventually enter the sealed enclosure and the arc at the contacts cause ignition. This is highly unlikely in a Division 2 location because of the relatively short time that flammable atmospheres are likely to be present. Also, even if the hermetic seal is not completely effective, the leak rate is still quite slow.

In Class I, Division 2 locations, as in Division 1 locations, attention must be given to surface temperatures. In Division 2 locations, however, only the normal operating temperature conditions of the equipment needs to be considered. One need not be concerned with high surface temperatures because of a stalled rotor on a motor or overheating of a connection point of two conductors, or even a heavily overloaded conductor with the resulting high surface temperatures. The abnormal conditions of the occurrence of a flammable atmosphere and an abnormal equipment fault condition are very unlikely to exist at the same time.

4-1.1.3 Zone 0

Since Zone 0 locations are those where the atmosphere is within the flammable range all or almost all the time under normal conditions (see definition in Section 2-4.3.3.1), the equipment design concept is that there must be at least two simultaneous faults in the protection system (three in intrinsically safe systems) before ignition can occur. Consideration is being given to increasing the level of safety so that at least three simultaneous faults will be necessary before ignition can occur for all protection systems in Zone 0, not just for intrinsically safe systems.

Explosionproof systems, even though acceptable for Division 1 locations, do not meet this criteria. A single fault, such as a loose or missing enclosure bolt or a bad scratch on a flat joint surface, can render the protection useless. The *National Electrical Code*[1] includes information in a fine-print note following the definition of Class I, Division 1 locations on the need for extra-special precautions in the most hazardous parts of Division 1 locations, those parts corresponding to Zone 0 locations.

When adopting the Zone system of area classification as an alternate to the long-standing Division system for the 1996 *NEC*, the National Electrical Code Committee, Panel 14, recognized that if some protection systems suitable for Division 1 locations were used in locations meeting the definition of Zone 0 but defined as Division 1, the protection would not be at as high a level as required of equipment in Zone 0 locations. The panel judged this an acceptable situation for several reasons. Such equipment (usually explosionproof equipment) had an excellent safety record over many years of service, so there was no good technical justification for a complete change in the classification system. There are safety factors built into the requirements for explosionproof equipment that exceed the safety factors for similar-appearing flameproof equipment. Locations where flammables within the explosive range exist all or most of the time are rare. They are estimated by some to be about 1 percent of Division 1 locations. In such locations, electrical equipment, if any, is usually instrumentation that is or can easily be made intrinsically safe, a design suitable for Zone 0 locations.

4-1.1.4 Zone 1

The design concept in Zone 1 is essentially the same as in Division 1, with some exceptions. Since a Zone 1 location is not actually hazardous all or even most of the time under normal conditions, the requirements need not be as stringent as in Zone 0 or Division 1. The safety factors built into the equipment design and testing need not, therefore, be as great in Zone 1.

This permits equipment designs not acceptable in Division 1 locations to be considered acceptable in Zone 1. The major ones are increased safety equipment, flameproof equipment, and a version of intrinsically safe equipment. Also acceptable in Zone 1 are oil-immersed, powder-filled, encapsulated, and a version of purged and pressurized equipment.

4-1.1.5 Zone 2

The design concepts for equipment for use in Zone 2 locations are the same as those for Division 2 locations. Some designs not previously recognized in the United States have been used internationally for many years, and these are expected to find a place in the United States under the requirements for Zone 2 locations.

4-1.2 Class II Locations

4-1.2.1 Division 1

The concepts for protection systems in Class II locations are similar to those in Class I locations. However, significant differences exist between Class I, Divisions 1 and 2 as compared to Class II, Divisions 1 and 2. In a Class II, Division 1 location, a dust cloud with a relatively high concentration of dust is anticipated under normal operating conditions, as is a thick layer of dust on the equipment. In Class II, Division 2 locations, the thick dust cloud is not present under normal conditions, and it is anticipated that the layer thicknesses will be considerably less in Division 2 than in Division 1 locations. Since it is not anticipated that an ignitible dust cloud will be present in a Division 2 location, although there may be some dust present in the air, ordinary dusttight equipment usually provides the necessary protection.

Groups: The group in which dust is included does not have the same significance in Class II locations that it does in Class I locations, but it does have some degree of significance. If Group E dusts are present, the *National Electrical Code* does not permit classification of the area as Class II, Division 2. (See Section 2-5.4.2.) Thus, all equipment in Class II, Group E locations is required to be suitable for Division 1 locations.

In general, dust blanketing temperature tests and dust penetration tests for Groups F and G are conducted using the same type of dust, but a different type of dust is used when testing for Group E locations. Except for the role of temperature as noted below, equipment suitable for Group G locations is usually considered suitable for Group F locations, and vice versa.

These group concepts are generally applicable regardless of the protection system used.

Temperature: The surface temperature of equipment is more critical in Class II locations than it is in Class I locations, because the safety factors associated with ignition temperature as determined for gases and vapors do not exist, at least to the same extent, for Class II locations. (See Section 2-5.) In addition, dust acts as a thermal insulator; that is, dust collected on electrical equipment reduces the effectiveness of the heat transfer from the heated electrical parts inside the enclosure to the air surrounding the enclosure. Equipment with a dust layer can be expected to operate at higher temperatures than the same equipment without a dust layer. Furthermore, the ignition temperature of a dust layer is usually lower than the ignition temperature of gases and vapors. (See Appendix A, Tables A-5 and A-6.) Therefore, for any specific piece of equipment there is a greater likelihood that the surface temperature will exceed the ignition temperature of a dust layer than that it will exceed the ignition temperature of a flammable gas or vapor. Ignition of dust layers by overheated equipment and fires and explosions resulting from this ignition are far more common in Class II locations than in Class I locations. It is therefore essential that careful attention be paid to the surface temperature of the equipment, including the surface temperature with a layer of dust when the equipment is used in a Class II, Division 1 location. When testing equipment for Class II, Division 1 locations, the surface temperature is measured when the equipment is coated with a heavy layer of dust.

4-1.2.2 Division 2

The only problem of major significance in a Class II, Division 2 location is light or occasional dust in the air and some dust accumulation on equipment. Protection systems in Class II, Division 2 locations normally consist of dusttight enclosures, such as NEMA 4. See NEMA Standards Publication No. 250 for complete information on NEMA enclosure types.[2] The major exceptions are motors and electric lighting fixtures.

For motors, totally enclosed motors, either of the nonventilated, fan cooled, or totally enclosed pipe ventilated types, are normally required.

Although the *National Electrical Code* permits fixed lighting equipment to be of other than the type listed specifically for use in hazardous locations, it is common practice to use lighting fixtures specifically listed by one of the various approval agencies for Class II, Division 2.

This is due to the problem of determining the external surface temperature of incandescent and high-intensity discharge (such as mercury vapor) lighting fixtures that are not marked to indicate the external surface temperature.

Groups: Except for the problem of external surface temperature, the group identification in Class II, Division 2 locations has little significance. As already noted in Section 4-1.2.1, there are no Division 2 locations where Group E (electrically conductive) dust is present.

Temperature: The temperature under normal operating conditions is considered the equipment's external surface temperature when testing the equipment for use in Class II, Division 2 locations.

4-1.3 Class III Locations

In general, if the equipment is suitable for Class II, Division 2 locations, it is also acceptable for Class III locations, provided the external surface temperature does not exceed 165°C (329°F). The requirements are essentially the same for either Division 1 or Division 2 in Class III locations.

4-2 Class I Locations

4-2.1 Division 1

4-2.1.1 Explosionproof Equipment

The most common and easily recognizable protection method for electrical equipment in hazardous locations classified under the requirements of Article 500 of the *NEC* (the "division" system) is equipment designed to be explosionproof. Explosionproof apparatus is defined in the *National Electrical Code* as the following:

> Apparatus enclosed in a case that is capable of withstanding an explosion of a specified gas or vapor which may occur within it and of preventing the ignition of a specified gas or vapor surrounding the enclosure by sparks, flashes, or explosion of the gas or vapor within, and which operates at such an external temperature that a surrounding flammable atmosphere will not be ignited thereby.

This definition includes a number of criteria. First, it obviously requires that the electrical apparatus be enclosed in a case (the explosionproof enclosure).

Second, it requires that this enclosure be capable of withstanding an explosion that may occur within it. The intent is that the enclosure have sufficient strength to withstand the pressures generated by an explosion, assuming that the flammable gas or vapor enters the enclosure. Because explosionproof enclosures have joints between their various parts to permit installation and conductor entry and to accommodate operators or shafts, it can be expected that flammable gases and vapors will enter the enclosure through these joints. In accordance with sound engineering treatment of mechanical parts, a safety factor is applied to this strength requirement to cover such variables as variations in material and material thickness during the process of manufacture, deterioration due to corrosion, and unusual conditions that cannot be covered by explosion tests.

The third criterion of the definition is that the explosion be of a specified gas or vapor. This ties back into the separation of Class I locations in the division system of classification into four groups (A, B, C, and D). It would be prohibitively expensive if explosion tests were conducted with every gas and/or vapor classified in a particular group. Therefore, specific gases or vapors (usually only one per group) are used to test explosionproof apparatus.

The fourth criterion is to prevent the ignition of the gas or vapor surrounding the enclosure by sparks, flashes, or explosion of the gas or vapor within the enclosure. The intent is that the joints between various parts of the enclosure, and between the enclosure and the connecting conduit or cable, be such that under explosion conditions, the explosion within the enclosure will not be propagated through the joint to the surrounding atmosphere.

The fifth and last criterion is that the equipment operate at such an external temperature that a surrounding flammable atmosphere will not be ignited by the outside of the enclosure. If an electric lighting fixture operates at such a high temperature that the external surface of the explosionproof enclosure around the lamp bulb exceeds the ignition temperature of the flammable atmosphere, the external surface of the enclosure itself can become the ignition source, defeating the purpose of the enclosure.

There are a number of different standards for explosionproof equipment, many of them recognized as American National Standards. Most of the standards giving detailed construction and performance requirements are published by Underwriters Laboratories Inc.

Types of Joints: Minimum construction requirements have been established by industry agreement for a number of different types of joints. This does not mean that a joint in an enclosure (other than the joint between the threaded conduit wiring system and the enclosure) will meet all of the applicable requirements, even if it meets the minimum construction requirements.

The effectiveness of a joint depends upon a number of factors. Experience has shown that a flat joint meeting the minimum construction requirements in the published standards will not necessarily meet the explosion test requirements. The bolts may not be strong enough or spaced closely enough, or the enclosure may not be stiff enough. This is particularly true for equipment intended for Group B and Group A locations. There are few, if any, enclosures with other than threaded joints suitable for Group A locations. Also, manufacturers have found that it is impractical to build electric motors and generators for use in Group B and Group A locations, not only because of the relatively small demand for such equipment, but because maintaining the extremely close tolerances between the rotating shaft and the housing is impractical on a production basis.

Flat Joints: Flat joints are the most directly related to maximum experimental safe gap (MESG) experiments and tests (see Section 2-4.2.3), and they are also one of the most common types of joints. (See Figures 4-1 and 4-2.)

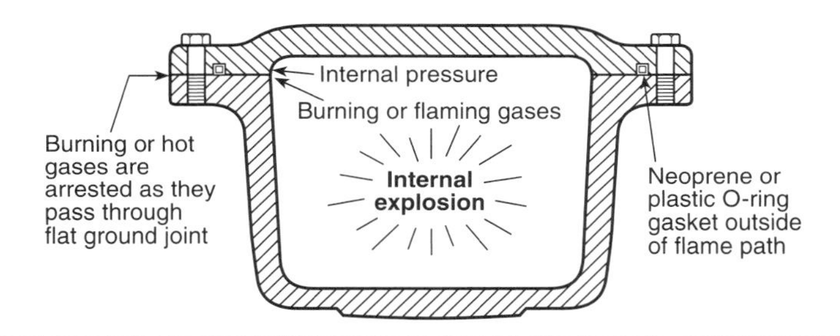

Figure 4-1. *A flat joint is commonly used for large enclosures and where threaded joints are impractical, such as for covers for rectangular enclosures. (Courtesy of HEP–Killark.)*

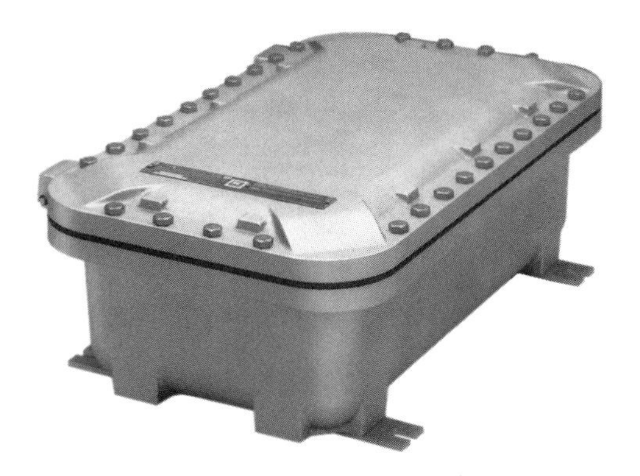

Figure 4-2. *Cast junction box with flat bolted joints. (Courtesy of HEP–Killark.)*

The construction requirements for flat joints in explosionproof enclosures are based on a minimum width of joint of 0.75 in. (19.1 mm) and a maximum clearance between joint surfaces of 0.0015 in. (0.038 mm). For enclosures with smaller volumes, particularly if the equipment is limited to Group D locations, the minimum width requirements may be reduced to 0.5 in. (12.7 mm), 0.375 in. (9.52 mm), or even 0.25 in. (6.35 mm). A bolt is permitted in the joint width if the distance from the inside of the enclosure to the nearest edge of the clearance hole for the bolt meets a minimum requirement, such as 0.5 in. (12.7 mm) for a 0.75-in. (19.1-mm) or wider joint.

The 0.0015-in. (0.038-mm) maximum clearance between joint surfaces is not based on the MESG for any particular material; it is based on industry agreement and the minimum practical clearance that can be maintained in production. The joint clearance is normally measured with a feeler gauge 0.125 in. to 0.5 in. (3.2 to 12.7 mm) wide, with a 0.5-in. (12.7-mm) wide gauge used, if possible.

A joint surface is required, when new, to have an arithmetical average roughness of not more than 250 microinches (0.0064 mm), in accordance with ANSI B46.1[3]. Standard roughness gauges can be used to make this determination, comparing the joint surface to the roughness on the gauge. A roughness of 250 microinches (0.0064 mm) can be maintained with ordinary machining and grinding operations. Polishing is unnecessary and may even reduce the effectiveness of the joint.

Threaded Joints: Another extremely common type of joint is the threaded joint. This is the joint used in conduit wiring systems. It is also commonly used for round covers and plugs, such as drain plugs. A threaded joint is probably the most effective type of joint, because it does not open up under explosion conditions, and because it presents an extremely long flame path. To be effective, however, the threaded joint must be wrenchtight. Although the minimum number of threads permitted in most standards is five fully engaged threads, the minimum number may increase if there are more than 20 threads per inch (25.4 mm). (See Figures 4-3 and 4-4.)

Rabbet Joints: A rabbet joint is, in effect, a flat joint with a right-angle bend in it. (See Figure 4-5.)

The "A" dimension shown in Figure 4-5 is known as the axial section of the joint, and the "B" dimension is known as the radial section of the joint. The joint clearance in the axial section is identified as the diametrical clearance, which is actually the difference between the inside diameter or dimension and outside diameter or dimension of the two mating parts of the joint. The clearance in the radial section of the joint is treated in a manner similar to the clearance between mating parts of a flat joint.

A rabbet joint has two major advantages. Since it is a right-angle

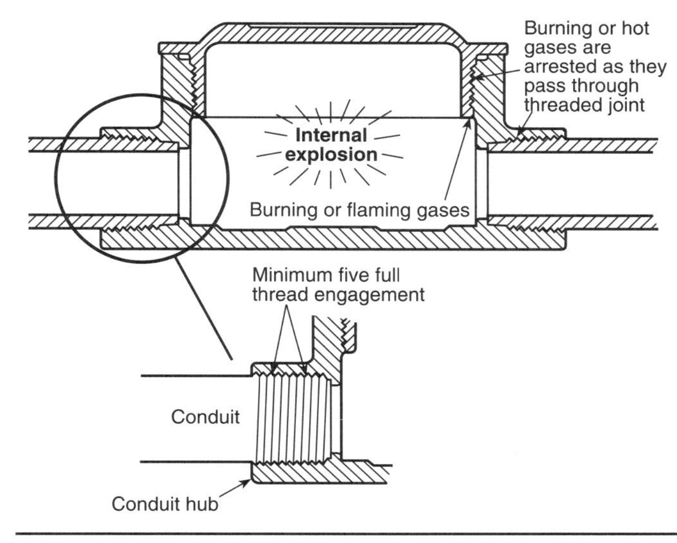

Figure 4-3. *A threaded joint with a minimum of five fully engaged threads. (Courtesy of HEP–Killark.)*

Figure 4-4. *Various types of threaded joints. (Courtesy of Appleton Electric Co.)*

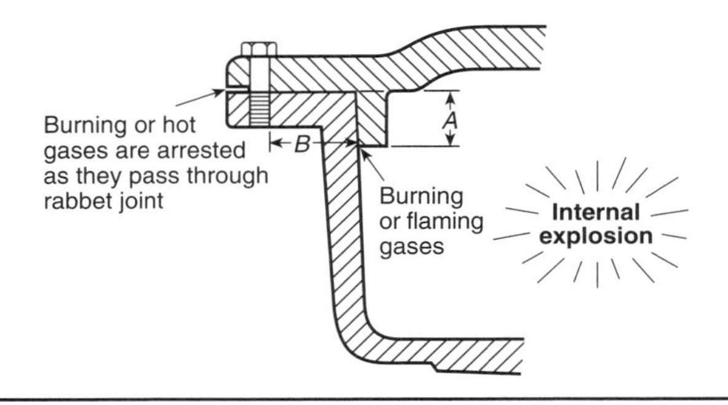

Figure 4-5. A rabbet joint has two dimensions, A and B. (Courtesy of HEP–Killark.)

joint it offers greater resistance to the flow of hot gases than an ordinary flat joint. In addition, it is ideally suited for a cylindrical fit, such as between a motor housing and the motor end bell. (See Figure 4-6.)

Labyrinth Joints: A labyrinth joint is, in effect, a series of rabbet joints. This type of joint approaches a threaded joint in effectiveness. (See Figure 4-7.)

A labyrinth joint is useful for small parts, such as slow-speed rotating shafts where a regular threaded joint cannot be used and other small parts where the advantages in the joint effectiveness outweigh the difficulty in manufacturing the joint.

Shaft Joints: A shaft joint is used wherever a high-speed rotating shaft must pass through the enclosure, typically in motors and generators but often for other moving parts. (See Figures 4-8 and 4-9.)

The requirements for joints for high-speed (100 rpm or more) rotat-

Figure 4-6. Rabbet joint at motor end bell.

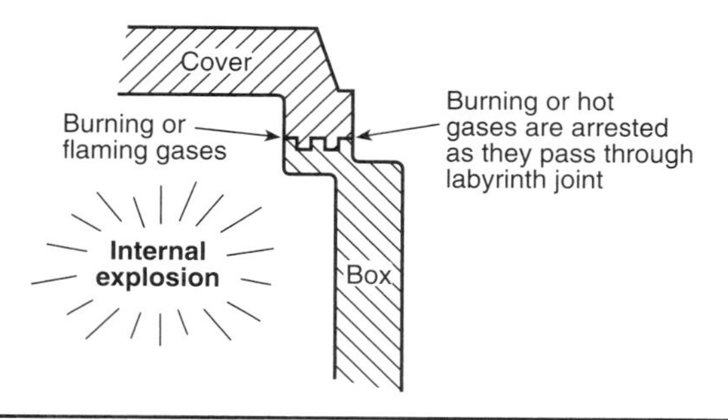

Figure 4-7. *Labyrinth joint. (Courtesy of HEP–Killark.)*

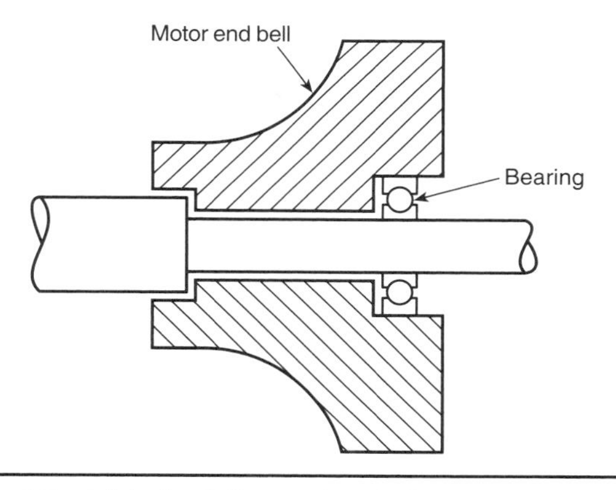

Figure 4-8. *Joint at rotating motor shaft.*

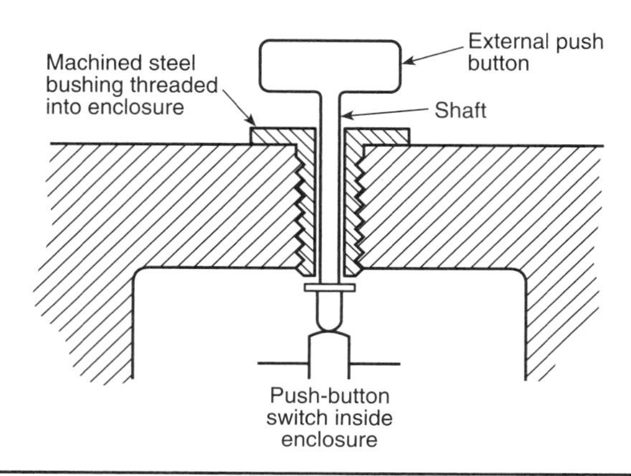

Figure 4-9. *Push rod for explosionproof push-button station.*

ing shafts permit larger clearances between the shaft and the housing than permitted for flat joints. The intent is to prevent contact between the two parts and wear at the joint. However, this requires that the length of the shaft joint (width in terms of flat joints) be greater than the equivalent flat joint. Since a shaft joint is inherently subject to potential wear, it should be checked periodically after the equipment is installed. (See Figure 4-10.)

Nonmetallic Materials: Although most explosionproof enclosures are made of metal, nonmetallic materials offer a number of advantages in industrial applications because of their high resistance to corrosion and light weight. Glass has been used in explosionproof enclosures for many years (in lighting fixtures and for small viewing windows), but polymeric (plastic) materials present several problems to equipment designers where they are used for explosionproof enclosures. Such problems include maintaining grounding continuity and resistance to solvent attack. Although metals are also subject to solvent attack and corrosion, the

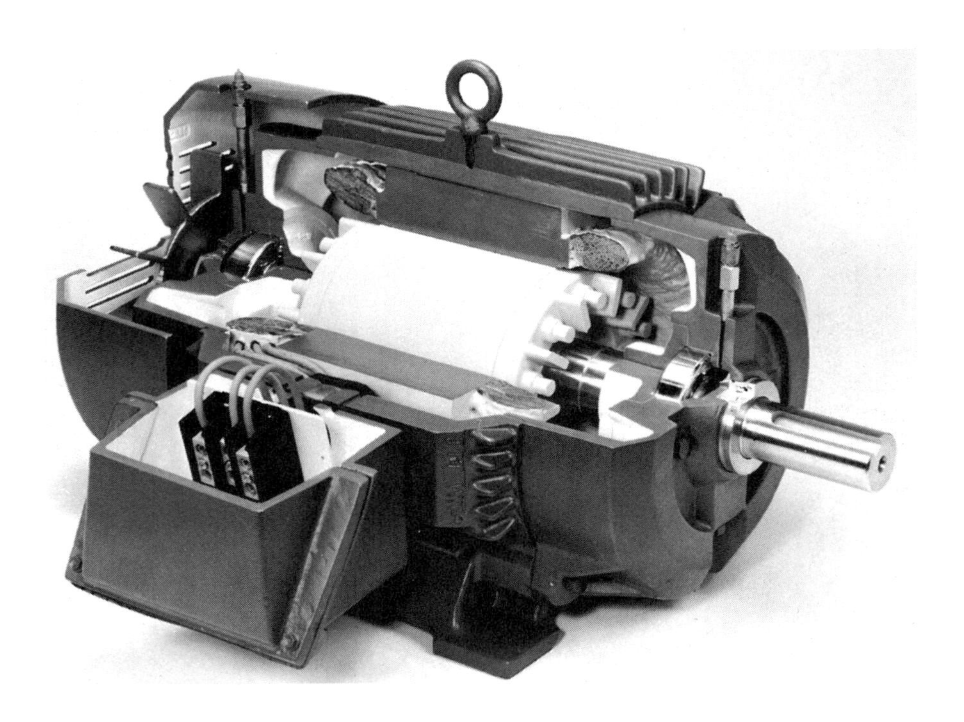

Figure 4-10. *Motor shaft joint. The motor bearings may be inside, outside, or both. If the bearings become worn, the shaft can also become worn, and the effectiveness of the joint in preventing propagation of an explosion can be reduced or destroyed. (Courtesy of Reliance Electric.)*

effects of such deterioration are quite evident visually. This is not necessarily true with polymeric materials. They may lose their strength with no visual evidence of deterioration.

More important, however, is the fact that joints made of polymeric materials behave differently than metal-to-metal or metal-to-glass joints. The polymeric material, unlike metal or glass, tends to wear or corrode away when explosion tests are conducted, reducing the effectiveness of the joint over a period of time. This phenomenon occurs even for the very high-grade glass-fiber-reinforced polyester materials. Because of this joint surface deterioration problem under explosion conditions, polymeric material joints in explosionproof enclosures are normally limited to threaded and labyrinth joints, where the effect is less pronounced.

Gaskets: Gaskets are not usually employed in explosionproof joints, but they may be used adjacent to a joint if they do not decrease the effectiveness of the joint. For example, an O-ring gasket adjacent to a flat joint may be used to make the enclosure weatherproof. (See Figures 4-1 and 4-11.)

Gaskets are often used where one of the joint surfaces is glass. Metal-jacketed gaskets, polytetrafluoroethylene (Teflon) gaskets, and gaskets of other materials specifically investigated for the purpose are used to prevent glass breakage.

Joint Changes under Dynamic Conditions: When an explosion occurs inside an explosionproof enclosure, the shape of the enclosure and the clearance between joint surfaces may be affected as a result of the pressures generated by the explosion. (See Figure 4-12.)

For a given enclosure, the higher the explosion pressure, the greater the change to the enclosure. The division system grouping of Class I materials into four groups (A, B, C, D) identifies not only the differences among the maximum experimental safe gaps of the various materials

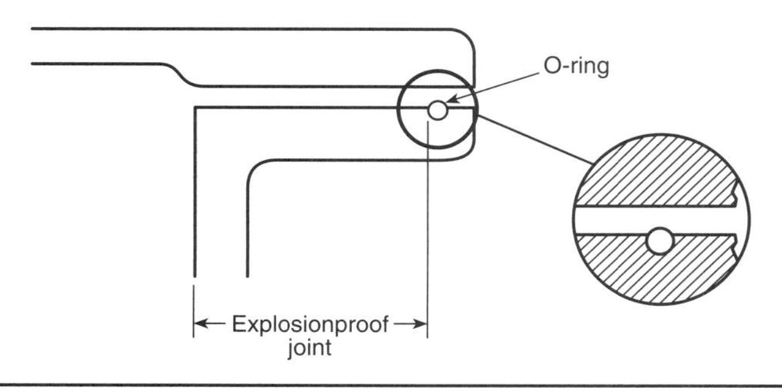

Figure 4-11. *An O-ring gasket on the outside of a joint, when properly designed, does not interfere with the performance of the explosionproof joint.*

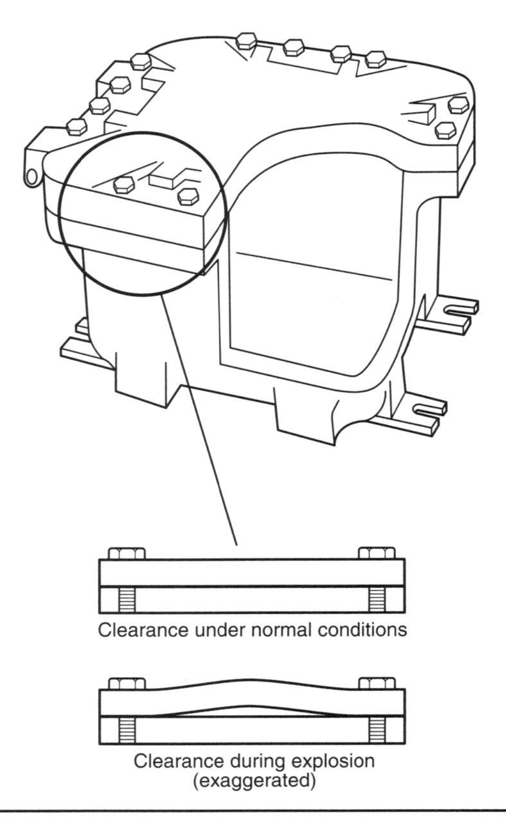

Figure 4-12. *The joint clearance changes under explosion conditions. (Courtesy of HEP–Killark.)*

but also the explosion pressures generated. The major reason Group A (acetylene) is separate from Group B materials (such as hydrogen) is that the pressures generated when acetylene-air mixtures explode are greater than the pressures generated when hydrogen-air mixtures explode. Even though the two materials have approximately the same maximum experimental safe gaps, propagation of the explosion through a flat joint is more likely in an acetylene-air explosion than in a hydrogen-air explosion, because the gap between joint surfaces is likely to open wider under the dynamic conditions of the explosion.

A number of factors affect the changes in a joint between two mating surfaces that are likely to occur as a result of the dynamic conditions of an explosion. Included are the stiffness of the enclosure itself, resulting from its shape, material, thickness, and method of fabrication; the mass of the enclosure; the rate of explosion pressure rise; and the material and size of bolts holding the enclosure together, as well as the spacing between bolts. For a typical flat bolted joint, higher-strength bolts and closer spacing between bolts may be needed for an enclosure designed for Group B or C locations than are needed for an enclosure designed for Group D locations. Higher strength can be achieved either by increasing the diameter of the bolt or the tensile strength of the material used. In addition to increasing the thickness or changing materials, ribbing and other reinforcement methods can be used to increase the stiffness of the enclosure at critical points.

Strength of Enclosure: The more pressure generated during an explosion, the greater the enclosure's strength must be to withstand the pressure without permanent deformation. Permanent deformation could be as severe as actual rupturing of the enclosure or as subtle as a change in the length of a bolt holding the enclosure together, because the elastic limit of the bolt was exceeded under the dynamic forces of the explosion.

The testing procedures for explosionproof enclosures are designed to result in the highest pressure likely under explosion conditions, taking into account both the group of hazardous location for which the enclosure is designed and the wiring requirements in the *National Electrical Code*. This includes consideration of pressure-piling, as described in Section 2-4.2.3. The *NEC* requirements for location of sealing fittings for the particular product being tested are also given consideration. For example, if a product under test is a switch enclosure, which is required by the *NEC* to have a sealing fitting located in each conduit entering the enclosure within 18 in. (457 mm) of the enclosure, tests are conducted with an 18-in. (457-mm) length of conduit attached. If the product is an outlet box with conduit openings not larger than 1½ in. trade size, tests are conducted with long lengths of conduit attached and with ignition at the ends of the conduit lengths to represent likely pressure-piling conditions

Explosion pressures increase as the energy of the ignition source increases. Therefore, if the enclosure is intended to contain a switching device that can be expected to interrupt high levels of current, such as a motor controller or circuit breaker, explosion tests are conducted with ignition by this high-energy source. Tests under this condition may result in higher explosion pressures than recorded when ignition is by a spark plug, which is the more common method of initiating an explosion when conducting explosion tests.

Extremely low ambient temperature (Arctic) conditions can also result in higher explosion pressures. Although extremely low ambient temperatures may reduce the likelihood of explosions because such temperatures are below the flash point for some flammable liquids, this is not true for flammable gases and for low-flash-point flammable liquids such as gasoline. Very low ambient temperatures often reduce the strength of some enclosure materials, so that the combination of reduced enclosure strength and increased explosion pressure can result in unanticipated problems with explosionproof enclosures used in low ambients. In addition, some sealing materials may not perform their intended function at very low temperatures.

There is considerable experience in the United States with explosionproof enclosures at low ambient temperatures in Group D atmospheres. Based on this experience and strength-of-material considerations, Underwriters Laboratories Inc. has indicated that explosion tests at ordinary room ambient temperatures represent explosion tests at temperatures as low as -13°F (-25°C) for Group D explosionproof equipment[4]. Operation of equipment below this temperature is also mentioned in the *NEC* as a potential problem requiring special consideration. If either lower ambient temperatures or extremely low ambient temperatures for other than Group D materials are anticipated, consideration should be given not only to the strength of the enclosure and effectiveness of seals at these low ambient temperatures but also to the likelihood of increased explosion pressures.

Some explosionproof motors have been tested and listed for ambient temperatures considerably below -13°F (-25°C). These listings are based on explosion tests and others conducted under low ambient conditions. The motors are marked with the minimum temperature for which they have been investigated.

External Surface Temperature: If the maximum operating temperature on the outside surface of equipment under either normal or anticipated abnormal conditions exceeds 212°F (100°C), the maximum operating temperature or temperature range is required to be marked on the explosionproof enclosure, together with other necessary identification, such as the hazardous location class and groups of materials for which the enclosure has been investigated. If the maximum operating temperature does not exceed 212°F (100°C) the equipment is considered, in effect, non-heat-producing, and is not required to be marked with the operating temperature or temperature range.

The conditions under which the temperatures are measured include both anticipated abnormal operating conditions and normal conditions. Abnormal operating conditions include single-phasing, overload, and stalled rotor for motors and blocked armature for solenoid valves. Normal conditions include rated voltage or rated wattage in any intended mounting position for lighting fixtures. Possible overheating due to other abnormal conditions, such as overcurrent conditions on current-carrying parts in switches and on conductors and arcing faults between conductors or to grounded parts, is intended to be reduced or alleviated by the *National Electrical Code* requirements for grounding and overcurrent protection.

Thickness of Enclosure: There are minimum thickness requirements specified for enclosures used for explosionproof equipment, based on the material used and the method of fabrication (casting, fabrication from sheet stock, and so on). Actual thicknesses may need to be greater than the specified minimums based on performance requirements. These minimum thickness requirements are intended to provide a degree of protection against burn-through or high external temperatures as a result of arcing faults to the enclosure. The material thickness is intended to be sufficient to provide time for the overcurrent devices in the circuit to open.

An enclosure is also intended to protect the internal wiring and live parts against damage due to likely mechanical abuse of the enclosure. Even though an enclosure may be of sufficient thickness to withstand the explosion pressures and is not subject to high-energy arcing faults [because it contains a Class 2 (low energy) circuit, for example], protection against mechanical damage is still necessary if the enclosure is to perform its intended function.

Wiring Methods: There are only a few wiring methods permitted for use in Class I, Division 1 hazardous locations, because protection of the energized conductors is just as important as protection of other energized parts. The most common wiring methods are threaded rigid metal conduit (either steel, aluminum, or bronze) and threaded steel intermediate metal conduit. A minimum of five fully engaged threads, made up wrenchtight, are required to provide adequate joints between parts of the conduit systems. Connections to enclosures should also be threaded: knockouts or their equivalent with locknut and bushing connections of conduit are not acceptable in Class I, Division 1 locations. The threaded couplings provided with these rigid conduit systems are fittings as defined in the *National Electrical Code*, even though usually they are provided as part of the conduit rather than as separate electrical fittings (such as capped elbows and unions). These conduit wiring systems are in themselves explosionproof enclosures, and their connection to an enclosure for splices, a switch, or other equipment extends the explosionproof enclosure to the point where a sealed conduit fitting for sealing is provided.

Explosionproof unions may be needed near sealing fittings and enclosures to assure that all threaded joints can be made wrenchtight. Ordinary conduit unions (Erickson couplings) should not be used, as the rotatable metal-to-metal joint inside such unions is not designed to be explosionproof.

Where flexibility is required, such as at motors where vibration can be a problem, the *Code* specifically permits flexible connection fittings approved for Class I locations. These fittings, too, are explosionproof. They are available in standardized lengths with a threaded metal fitting at each end. They consist of a seamless corrugated metal tube protected against mechanical damage by a metal braid on the outside. They also have a nonmetallic liner on the inside to reduce the likelihood of arcing to, and burn-through of, the seamless tube. (See Figure 4-13.)

Mineral-insulated, metal-sheathed (Type MI) cable is also recognized as an acceptable wiring method in Class I, Division 1 locations. This wiring consists of a somewhat pliable but not flexible cable made up of uninsulated conductors surrounded by compacted magnesium oxide and protected by a seamless metal tubing. Although this wiring method is not "explosionproof" by definition, the magnesium oxide acts as both an electrical insulator for the conductors and as a continuous seal. Ordinary Type MI cable fittings are not permitted in Class I, Division 1 hazardous locations. Special explosionproof Type MI cable fittings are necessary, primarily because the unthreaded joints between the outside of the cable and the inside of the fitting must be investigated as explosionproof. Also, the seals that prevent moisture from entering the cable need investigation to determine their resistance to the solvents likely to be encountered. (See Figure 4-14.)

Another wiring method has been permitted for the first time in the 1996 *NEC*. It is expected to find wide use, particularly in the petrochemi-

Figure 4-13. *An explosionproof flexible connection fitting.*

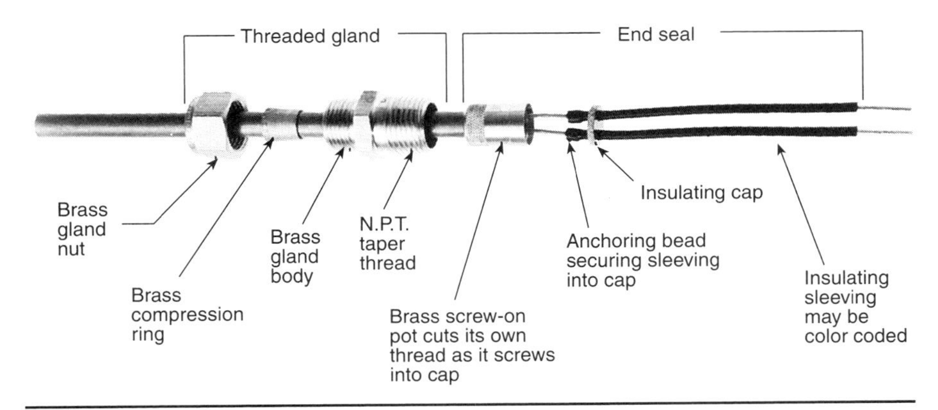

Figure 4-14. *Type MI cable and fitting. The flat joint between the gland body and the compression ring and between the compression ring and the gland nut must be explosionproof. Also, the threaded connections must be explosionproof. Type MI cable fittings not investigated for use in hazardous locations may not be explosionproof.*

cal industry where it has been used successfully for a number of years on offshore oil platforms. The new wiring method is not explosionproof and is not permitted where it may be subject to physical damage. It is a cable wiring method using a special Type MC cable with a continuous gas/vaportight corrugated aluminum sheath covered by an overall polymeric jacket. It is required to be listed for use in Class I, Division 1 locations. Its use is limited to industrial establishments with limited public access, where the conditions of maintenance and supervision ensure that only qualified persons will service the installation. The termination fittings used are required to be listed for the application. These fittings have provisions for sealing the end of the cable. (See Figure 4-15.)

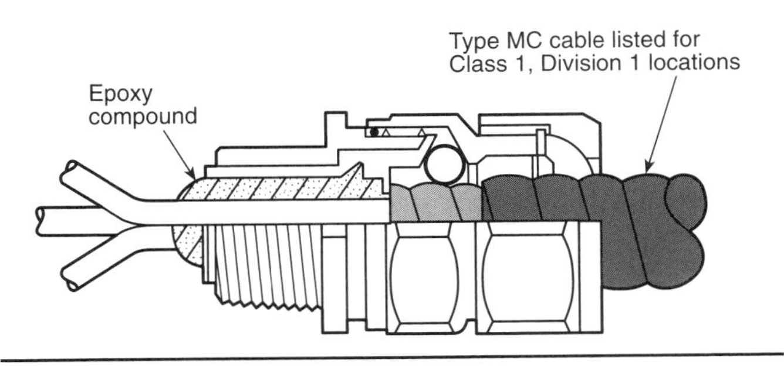

Figure 4-15. *Listed sealing fitting for Type MC cable in Class I, Division 1 locations. (Courtesy of HEP–Killark.)*

Another method also new in the 1996 *NEC* for general use in Division 1 locations is rigid nonmetallic conduit encased in a 2-in. (50.8-mm) envelope and buried at least 2 ft (610 mm) underground. (See Figure 4-16.)

A flexible cord is not a wiring method recognized in the *National Electrical Code*, although the *NEC* recognizes that it may be necessary to use a flexible cord for connection of portable equipment through use of attachment plugs and for short portions of the circuit in industrial establishments where fixed wiring methods will not provide the necessary degree of flexibility. A typical example is on an industrial robot. Where used, the cord is required to be of a type approved for extra-hard usage (Type S, SO, and so on, or Type W or G multiconductor cable). Since the cord is usually the weakest link in the protection system, its use is very limited. The *NEC* does not recognize explosionproof extension cords, for example. If used for portable equipment, the cord must be unbroken from the equipment to the fixed portion of the supply

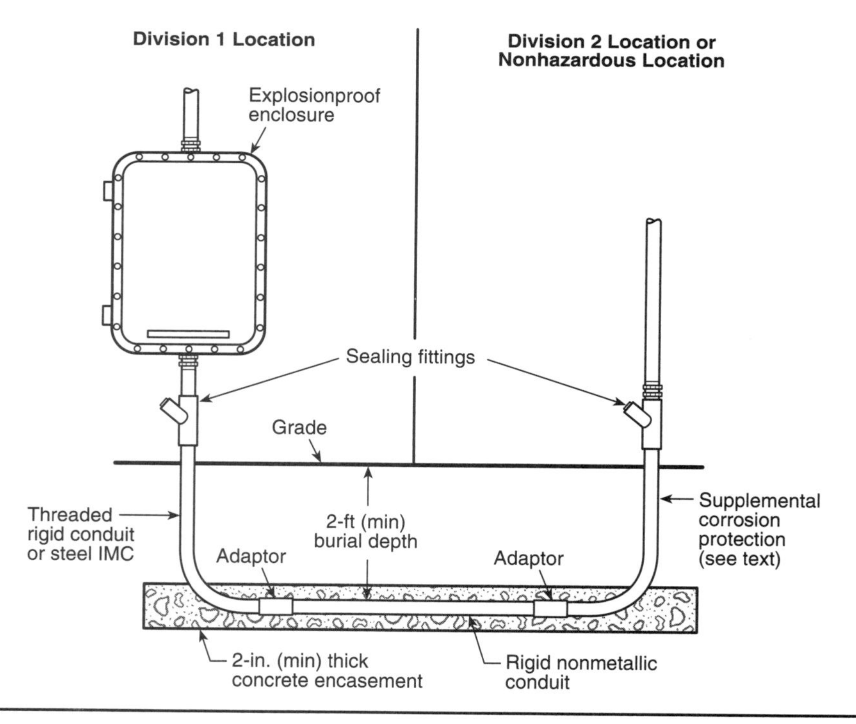

Figure 4-16. *Use of rigid nonmetallic conduit in Division 1 locations. Because severe corrosive effects are likely to occur on the metal wherever ferrous metal conduit runs directly from concrete encasement to soil burial, supplemental corrosion protection may be necessary on the conduit. An equipment grounding conductor will also be necessary in the conduit. (Courtesy of HEP–Killark.)*

circuit (the receptacle outlet). Every effort should be made to eliminate the use of flexible cord except where there is no possible substitute, such as for use in connecting portable equipment to the supply through an explosionproof plug and receptacle.

Other uses of flexible cord in Class I, Division 1 locations recognized in the *National Electrical Code* are for connection of electric submersible pumps with means for removal without entering the wet pit and for electric mixers intended for travel into and out of open-type mixing tanks or vats. In these cases, the *NEC* indicates that such equipment be considered portable utilization equipment.

In addition to the severe limitations on the use of flexible cord in Class I, Division 1 locations, the *NEC* warns against possible deterioration of the insulation by the flammable materials involved. Some cord components deteriorate very rapidly when exposed to the deleterious effects of flammable liquids, many of which are solvents.

Approval laboratories require that portable explosionproof equipment be provided with warnings about inspecting the cord and that the construction be such that the cord can be replaced easily without damage to the explosionproof enclosure. Special fittings are listed for this purpose. (See Figure 4-17.)

In Class I, Division 1 locations, all boxes and fittings are required to be suitable for, and marked to indicate their suitability for, the particular hazardous location class and group involved. This includes boxes, elbows, conduit bodies, and unions. The only exceptions are the threaded couplings provided as part of the rigid conduit or intermediate metal conduit. Also, conduit bends and threaded nipples made from conduit are permitted.

Seals: Seals serve several purposes in Class I, Division 1 locations. They are required in each conduit run leaving the Division 1 location, whether the conduit run is entering a Division 2 location or a nonhazardous location, to minimize the passage of gases and vapors through the conduit from the hazardous to the less-hazardous or nonhazardous location. Flammable gases can be transmitted from one location to another simply by pressure differentials at the ends of a horizontal conduit run as a result of air movement outside the conduit. Gases and vapors can move through nonhorizontal runs without any pressure differential. *NEC* requirements prevent the explosionproof protection system from changing the area classification, because the conduit acts as a "pipe" to transmit the flammable material from one location to another. (See Figures 4-18 and 4-19.)

Seals are required in Division 1 locations to complete the explosionproof enclosure. As has already been noted, the conduit is part of the explosionproof enclosure system, since it provides an entry into the explosionproof enclosure.

Seals are also required to prevent the passage of flame from one

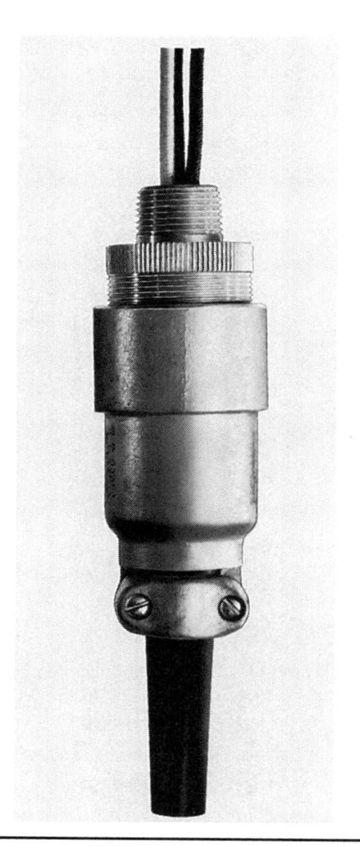

Figure 4-17. *Factory-sealed flexible cord connector. The flexible cord can be replaced without entering the explosionproof enclosure.*

portion of the electrical installation to another through the conduit system. In other words, seals help to prevent pressure-piling through the conduit system. Pressure-piling can greatly increase explosion pressures to the extent that enclosures can rupture or joints can open up to permit ignition of the surrounding flammable atmosphere.

The *NEC* requires that explosionproof enclosures housing parts that are ignition sources under normal operating conditions, such as switches, circuit breakers, and high-temperature parts, be sealed within 18 in. (457 mm) of the enclosure. It is assumed that under normal conditions flammable mixtures will enter the explosionproof enclosure, and if, also under normal conditions, the product enclosed is a source of ignition, the likelihood of pressure-piling from this enclosure to another through the conduit system is much greater than it would be if the enclosed

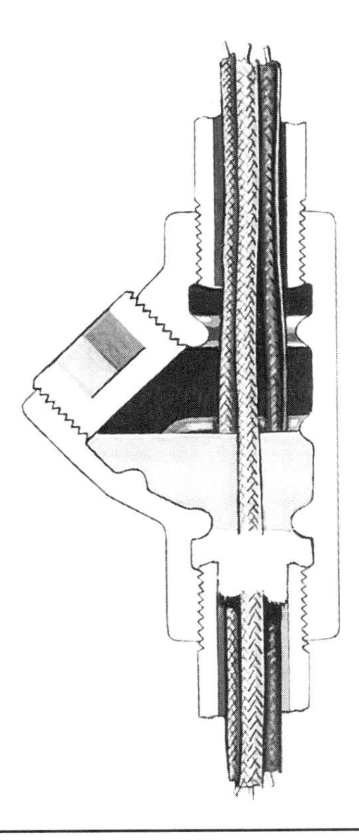

Figure 4-18. *A sealing fitting in a vertical run of conduit.*

device was not ignition-capable under normal conditions. (See Figure 4-20.)

The *NEC* also requires seals within 18 in. (457 mm) of each enclosure for conduits larger than $1^1/_2$ in. trade size. Put another way, if the explosionproof enclosure does not house an ignition-capable part, and the conduit system is $1^1/_2$ in. trade size or smaller, a seal is not required at the enclosure. This is partly a recognition of the increased likelihood of pressure-piling through the larger diameter conduits, the higher explosion pressures involved with larger enclosures, and a recognition of the economic factors involved in wiring in hazardous locations. Experience has shown that the rules in this regard provide adequate safety. (See Figures 4-21 through 4-27.)

Some flammable gases (butadiene, ethylene oxide, propylene oxide, and acrolein are known examples) have maximum experimental safe gaps putting them into one hazardous location group, but their explosion

Figure 4-19. A typical sealing fitting installation. Note the explosionproof conduit unions next to each sealing fitting to ensure wrenchtight connections. (Courtesy of HEP–Killark.)

pressures under pressure-piling conditions put them into a more severe group. (See Section 2-4.2.3.) Sealing all conduits, regardless of trade size and regardless of what is in the enclosure, will permit the use of more readily available and lower-cost explosionproof equipment than would be required if all conduits were either not sealed or were sealed in accordance with the minimum requirements in the *NEC*.

Testing Procedures: Three basic types of tests are conducted on explosionproof equipment, in addition to any tests required for the same equipment intended for use in nonhazardous locations. These are explosion tests, temperature tests, and hydrostatic pressure tests.

Explosion tests are conducted to determine the pressure to be used for the hydrostatic pressure test and to determine that the construction of the enclosure is such that it will truly prevent ignition under actual explosion conditions. For some types of products, such as circuit breakers, the tests are also conducted to determine that the enclosed device will perform its function under the unusual and severe conditions involved with flame and turbulence inside the enclosure.

Temperature tests are conducted to determine not only that the product meets ordinary location equipment requirements for tempera-

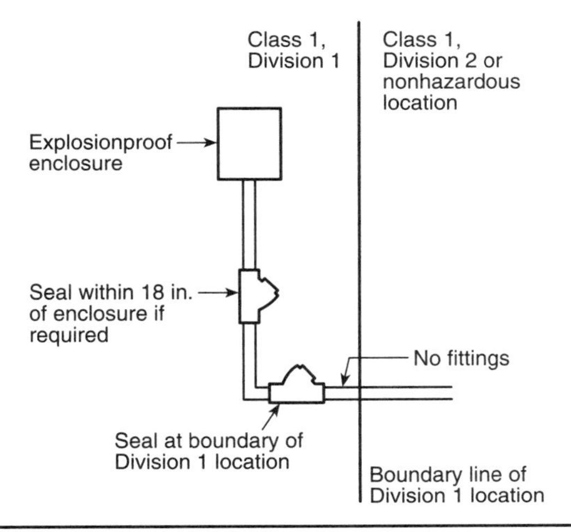

Figure 4-20. *The NEC requires a seal at the boundary of the Class I, Division 1 location. The seal can be on either side of the boundary and up to 10 ft (3.05 m) from the boundary if there is no box or fitting between the boundary and the sealing fitting. The seal must be designed and installed so as to minimize the amount of gas or vapor that can be communicated to the conduit on the Division 2 or nonhazardous location side of seal.*

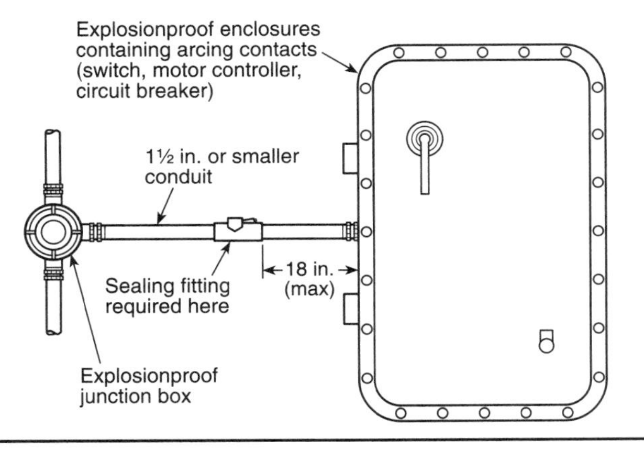

Figure 4-21. *NEC requires a seal in Class I, Division 1 locations if the enclosure contains equipment that under normal operating conditions can be an ignition source. (Courtesy of HEP–Killark.)*

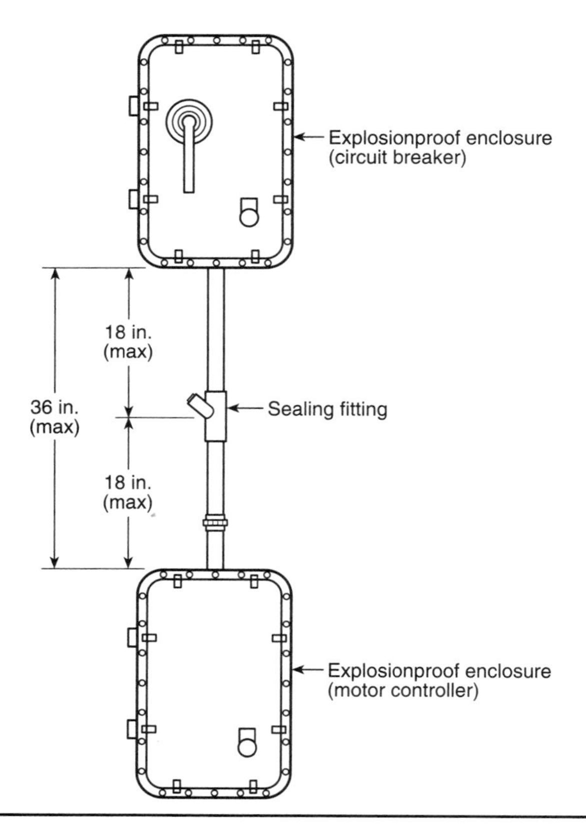

Figure 4-22. *NEC permits a single seal to serve two enclosures requiring sealing if the conduit run is not more than 36 in. (914 mm) long and the seal is not more than 18 in. (457 mm) from the enclosures requiring sealing. (Courtesy of HEP–Killark.)*

tures of the various parts but also to determine that the operating temperature marking, if required, is appropriate.

Hydrostatic pressure tests are conducted to demonstrate that the strength of the enclosure is adequate to withstand the explosion, with a factor of safety of usually four times the peak explosion pressure.

Explosion Tests: When conducting explosion tests, the flammable material used is dependent upon the hazardous location groups for which the enclosure is designed. Tests for Group D only are normally conducted using propane or, in some cases, pentane. Until the 1980s, Underwriters Laboratories Inc. conducted all Group D testing for explosionproof enclosures with a naphtha compound representing unleaded gasoline. This compound is still used for tests on flame arresters.

For enclosures intended for Group C locations, and for Groups

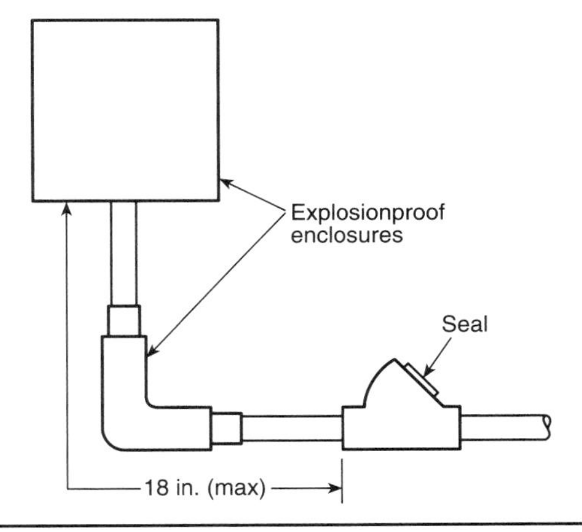

Figure 4-23. *The only enclosures permitted by the NEC between the sealing fitting and the enclosure being sealed are explosionproof unions, couplings (including threaded rigid conduit and intermediate steel conduit couplings), elbows, capped elbows, and conduit bodies similar to "L," "T," and "cross" types (conduit body of "L" type shown). The conduit bodies are not permitted to be larger than the largest trade size of conduit connected to the conduit body.*

C and D locations, the test gas is usually ethylene. Until the 1980s, Underwriters Laboratories Inc. normally used ethyl ether for this test. Explosion testing for Group C locations represents explosion testing for Group D.

For Group B locations, hydrogen is the test gas used. Explosion testing for Group B locations represents explosion testing for Groups C and D.

Equipment intended for Group A locations is tested using acetylene. Explosion testing using acetylene does not represent Group B, although it does represent Groups C and D. Even though the maximum experimental safe gaps are about the same and acetylene-air mixtures produce higher explosion pressures, hydrogen-air mixtures require less energy for ignition, and (thermal) energy released through the joints in the form of hot gases is a potential ignition source.

When conducting explosion tests, a series of tests are conducted over the critical range of explosive mixtures, normally around the stoichiometric mixture. The mixture that results in the highest pressure is not usually the same as the mixture that is the most easily propagated through a joint, and neither is usually the stoichiometric mixture. It is a combination of both factors that determines whether or not the enclosure is explosionproof. Also, there are variables in explosion testing that

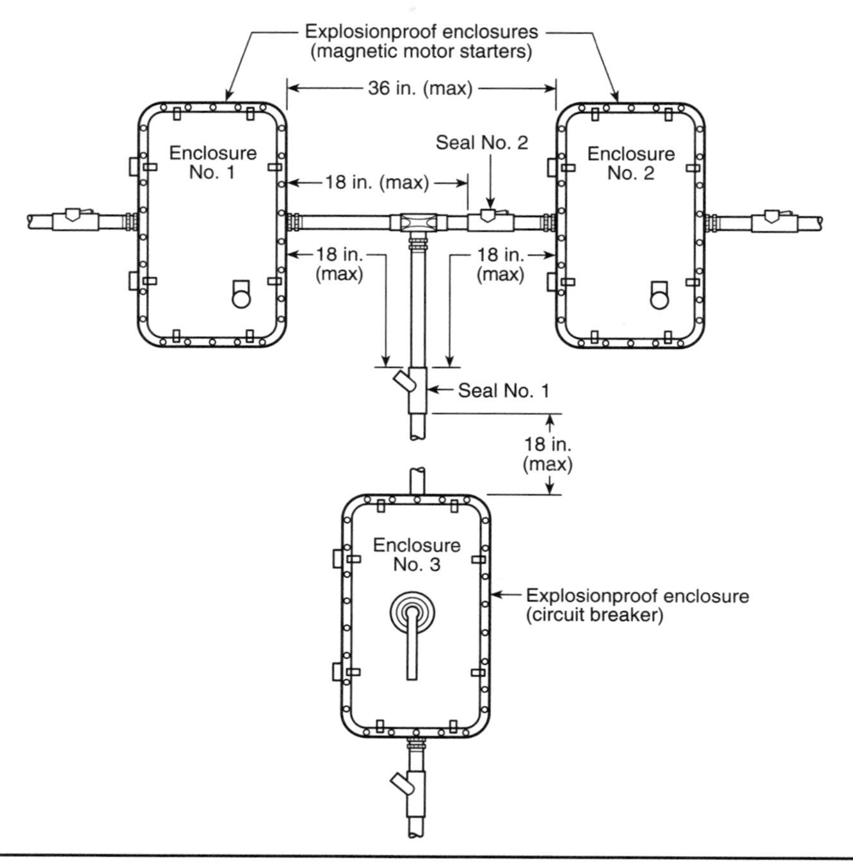

Figure 4-24. *In this case, the NEC requires two seals so that the run of conduit between Enclosure No. 1 and Enclosure No. 2 is sealed. Even if Enclosure No. 3 was not required to be sealed, the seal in the vertical run of conduit to Enclosure No. 3 would be required to be sealed within 18 in. (457 mm) of Enclosure No. 1, because the vertical conduit run to the "T" fitting is a conduit run to Enclosure No. 1. (Courtesy of HEP–Killark.)*

cannot easily be controlled, such as exactly how the hot gases ejected through enclosure joints mix with the unburned flammable mixture around the enclosure. A series of tests is therefore necessary to cover these variables.

Explosion tests are conducted with lengths of conduit attached and with ignition at the ends of the lengths of conduit to represent installation with or without conduit seals, depending upon the trade size of the conduit likely to be used with the equipment and the type of equipment to be enclosed. For example, a junction box for Group C locations intended to house splices or terminals with no conduit entry larger than

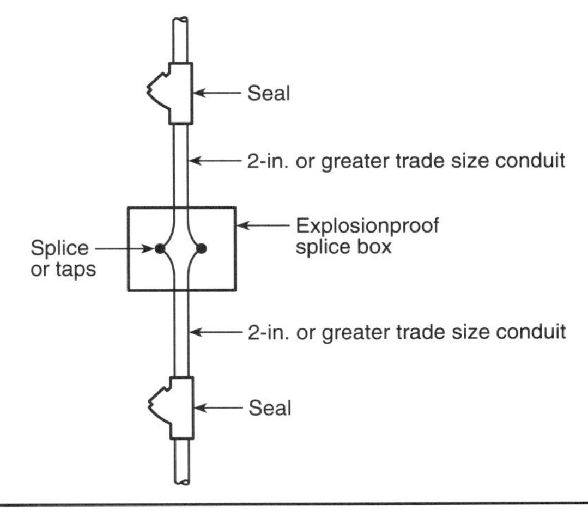

Figure 4-25. *In a Class I, Division 1 location, all conduits 2 in. trade size and larger are required to be sealed within 18 in. (457 mm) of any enclosure, regardless of what the enclosure contains.*

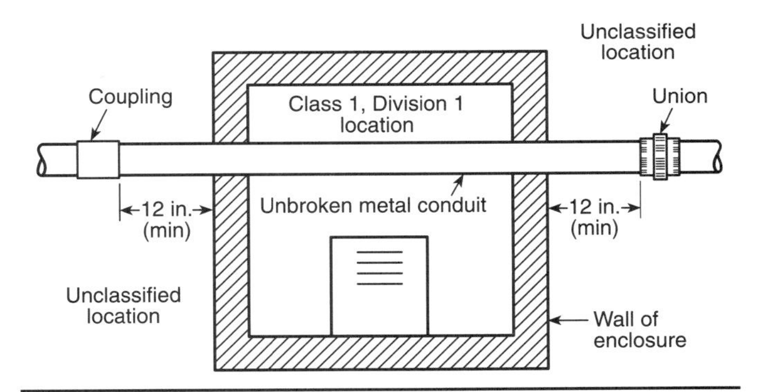

Figure 4-26. *Seals are not required by the NEC if the conduit passes completely through a Class I, Division 1 location with no fittings or boxes in the hazardous location or within 12 in. (305 mm) beyond the Division 1 boundary, and the conduit terminates in a nonhazardous location. Conduit runs may be vertical as well as horizontal or a combination of vertical and horizontal. (Courtesy of HEP–Killark.)*

$1^1/_2$ in. trade size would be tested with 5-, 10-, and 15-ft (1.52-, 3.05-, and 4.57-m) lengths of the largest trade-size conduit attached. If the conduit entries were larger than $1^1/_2$ in. trade size, the tests would be conducted with an 18-in. (457-mm) length of the largest trade-size conduit attached, as well as with long lengths of conduit if conduit entries $1^1/_2$ in. trade size and smaller are provided or if it is anticipated

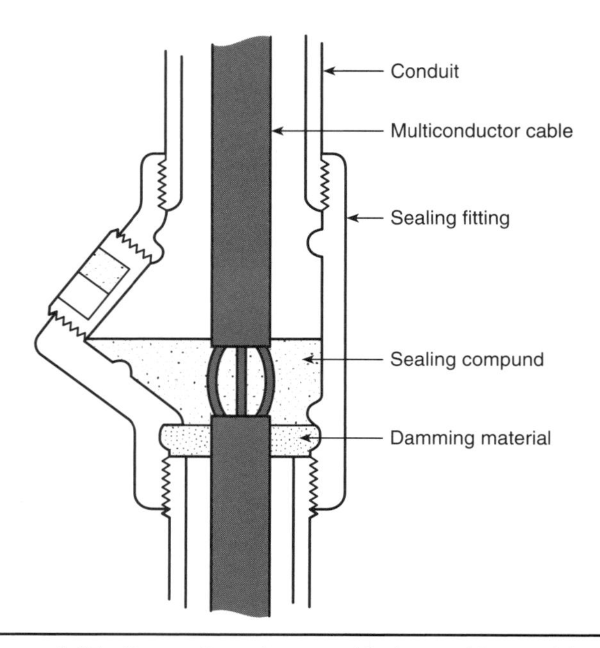

Figure 4-27. *If a multiconductor cable is used in conduit and the cable is capable of transmitting gases or vapors through the cable core, the cable jacket and coverings should be removed in the sealing fitting and the individual insulated conductors of the cable and the outer jacket should be sealed. (See Section 4-2.2.9 and Figure 4-45 for information on types of cables.) There is an exception for this removal of all coverings rule in the NEC for shielded cables and twisted pair cables. If the multiconductor cable is incapable of transmitting gases or vapors through the cable core, the entire cable can be considered a single insulated conductor and sealed accordingly.*

that smaller trade-size conduits will be used. Motor controllers and other switching devices for Group C locations would be tested with an 18-in. (457-mm) length of the largest trade-size conduit that is intended to be attached.

The point of ignition would also be varied based on the testing experience of the laboratory involved. In addition to ignition at the end of the length of conduit, tests would be conducted with ignition close to joints and with ignition by an arc at the contacts of the switching device itself.

If the equipment is intended to house a switching device, tests may be conducted with the switching device operated under anticipated load making and breaking conditions. For example, a motor controller would be tested with ignition resulting from a circuit designed to represent stalled rotor conditions of the motor. A circuit breaker would be tested using a circuit designed to represent a short circuit or ground fault.

It has been found that some devices designed and intended to interrupt large amounts of energy, such as motor controllers, large switches, and circuit breakers, will not always perform their intended function under the severe conditions of an explosion. An explosion involves flame, which ionizes the environment. Spacings between energized parts of opposite polarity and from energized parts to grounded parts may be inadequate for circuit interruption in an ionized medium, even though they are adequate under "normal" (nonhazardous location) conditions of operation.

In addition, arcing contacts may be enclosed as part of the construction of the product. Typical would be a molded-case circuit breaker mounted in an explosionproof enclosure. The pressures resulting from the explosion can cause the case of a molded-case circuit breaker to either implode or explode, depending upon the test conditions, if the molded plastic case does not have sufficient strength. Since a circuit breaker is a safety device, it is essential that it perform its function under the conditions likely to be encountered. Ordinary location testing does not cover this type of use.

Tests are also conducted using the arc at the contacts of switching devices as the ignition source, because explosion pressures normally increase with the higher energy arcs.

In some types of products, motors for example, pressure-piling can occur within the explosionproof device itself, with no relation to conduit connected to the enclosure. For example, passages through the motor from one end-bell to another can result in higher pressures at one end of the motor when the flammable mixture is ignited at the other end. The same is true for other types of products where the enclosure is divided into compartments by the product installed.

Turbulence normally results in higher explosion pressures than quiescent mixtures. Motors and generators, therefore, are tested under both running and nonrunning conditions.

When conducting the explosion tests, the product under test is placed in a chamber so that it can be surrounded by the flammable mixture. This chamber may be a specially constructed chamber designed for testing a large number and variety of devices, or it may be a chamber fabricated specifically to test the product under consideration. Some testing laboratories prefer to use permanent explosion testing chambers. Others prefer using test chambers specifically constructed for each product being tested, because the volume of gas and the hazards to test personnel and equipment can be reduced to a minimum should there be a test failure resulting in propagation of the explosion from inside to outside the enclosure under test. Under the latter conditions, the enclosure is usually made of wood with a taped-on plastic window that also serves as an explosion relief vent. The device under test is filled with a flammable mixture, either by passing the mixture first through the device under test and then into the surrounding chamber, or by piping the mixture into both the product under test and the test chamber at the same time. See Figures 4-28 through 4-31.

The flammable mixture to be used for the test is selected and piped into and out of the testing arrangement through a series of valves until the air or mixture from the preceding test has been completely purged and replaced with the flammable mixture to be used for the test. The valves are then closed and the mixture is ignited. Pressures are measured with transducers located at critical points.

Figure 4-28. *Explosion test setup at Underwriters Laboratories Inc.*

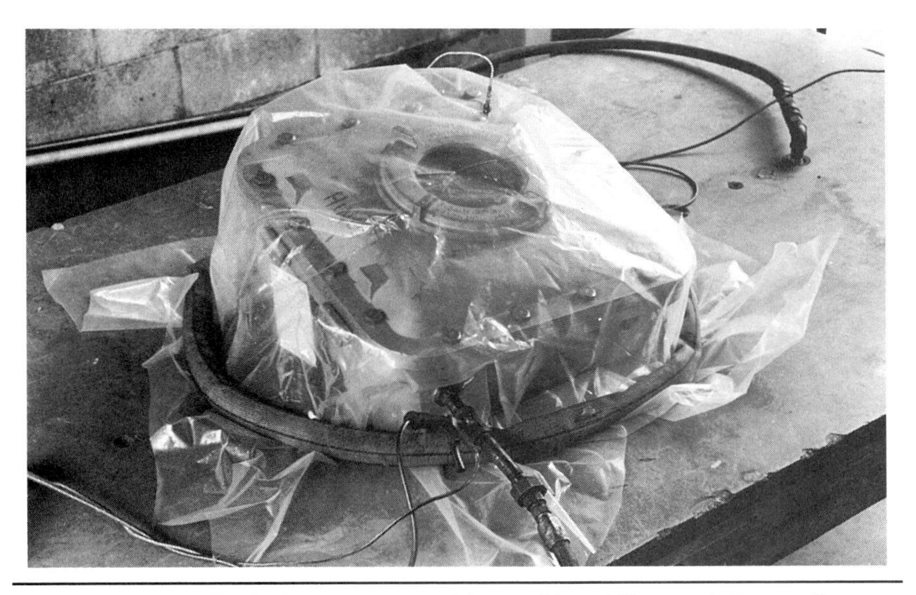

Figure 4-29. Explosion test setup at Factory Mutual Research Corporation.

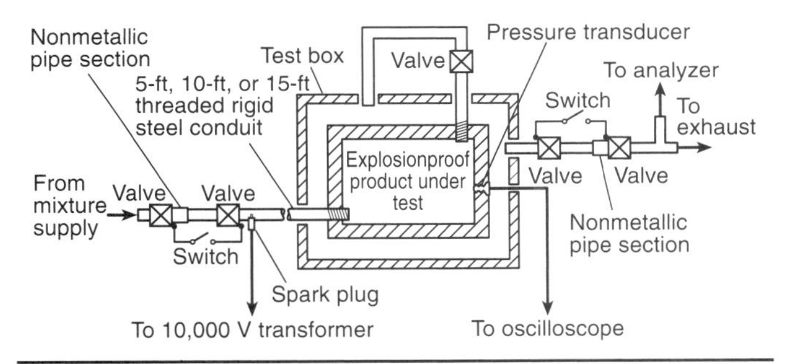

***Figure 4-30.** One piping arrangement for conducting explosion tests. In this arrangement, the explosive gas-air mixture flows through the enclosure under test then through the test chamber before going to the exhaust system. Flow is stopped and the enclosure is isolated from the test chamber by closing the valves before igniting the mixture. See also Figure 4-31.*

The test is repeated with a variety of flammable mixtures in the known critical range, under the various test conditions selected, until 10 to 20 tests have been completed, depending upon the variety of conditions involved.

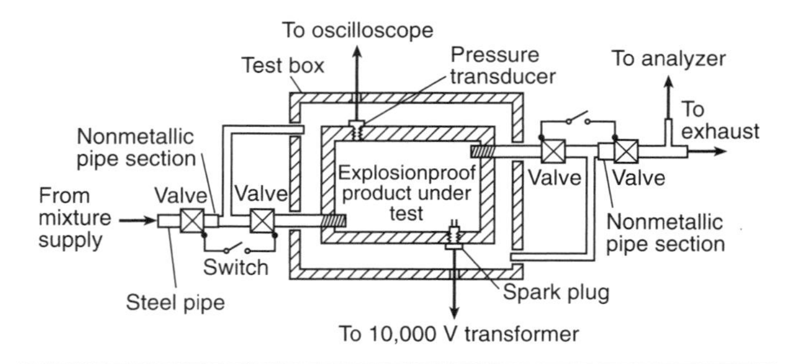

Figure 4-31. *Another piping arrangement for conducting explosion tests. In this case, a parallel piping arrangement is used. The nonmetallic pipe sections are used to isolate the equipment under test so that electrical tests can be conducted with the explosionproof enclosure connected to one side of the electrical supply line. The switches short-circuit this arrangement during filling and purging to prevent buildup of static charges. See also Figure 4-30.*

Temperature Tests: Temperature tests conducted on explosionproof equipment are essentially the same as temperature tests conducted on general-use (nonhazardous location) equipment, with two exceptions. First, temperatures are measured on the outside of the explosionproof enclosure to determine the external surface operating temperature in a 104°F (40°C) ambient. And second, tests are also conducted under anticipated abnormal conditions, such as overload, single-phasing, and stalled rotor for motors and blocked armature for solenoid valves.

Hydrostatic Pressure Tests: Hydrostatic pressure tests are conducted by filling the enclosure with water until all air has been eliminated, and then applying pressure at a rate of 100 to 600 psig (690 to 4137 kPa) per minute until the required internal pressure is reached. For safety, water (or other liquid) is used instead of air in case there is a test failure. It is often necessary to use gaskets or other means on the inside of the enclosure to prevent leakage of water during application of the pressure, depending upon the pumping capacity of the hydrostatic test equipment and the leakage rate of the enclosure under test. The enclosure is required to withstand the specified pressure for one minute, without rupture or permanent distortion.

In some cases it may be possible to calculate the strength of the enclosures. The maximum explosion pressure developed during the explosion test is used as a basis for determining the pressure to be applied, in accordance with Table 4-1.

Installation and Maintenance Precautions: As with any equipment, precautions are necessary during installation. The special protective

Table 4-1 Safety Factors for Determining the Strength of an Enclosure

Enclosure Material or Part	Safety Factor for Calculations	Safety Factor for Hydrostatic Pressure Tests
Cast metal	5	4
Fabricated steel	4	3*
Bolt	3	3

**The enclosure is required to withstand a hydrostatic pressure test of at least twice the maximum internal explosion pressure, without permanent distortion, and at least three times the maximum internal explosion pressure, without rupture.*

features should be treated carefully. Joint surfaces should be protected so that they are not scratched or damaged. If the equipment is to be painted after installation, care should be taken to ensure that the paint does not get on the joint surfaces.

Bolts and nuts should be drawn up tight, and if a specific torque is specified by the manufacturer, that torque should be used. Bolts left out or not tightened can reduce or destroy the effectiveness of the explosionproof enclosure. Similarly, conduit should be tightened wrenchtight, and threaded covers, plugs, and so forth should be tightened securely with the proper tool. Hand-tightening is not adequate.

Care should be taken to determine that the various components of the explosionproof system are compatible with each other. For example, conduit fittings for sealing are tested based on the use of specified sealing materials. Only these sealing materials should be used, and only in accordance with the instructions on the package. Unless otherwise specifically indicated in the instructions with the product, a sealing compound made by manufacturer A should not be used with a sealing fitting made by manufacturer B. Although it may appear to be the identical compound, sealing compounds have additives, often unique to the particular manufacturer, that cause the compound to expand during the drying process. This expansion can result in large mechanical forces on the fitting itself, and it could crack or otherwise damage the fitting if it is not designed for that particular sealing compound. In a similar vein, too little expansion can reduce the effectiveness of the seal.

Most sealing compounds for use in conduit sealing fittings are powders that must be mixed with water. Most indicate a temperature range over which they can be installed. Such instruction should be followed carefully so that proper setting of the compound is achieved.

Be sure that seals are actually poured. Some installers have been known to dribble sealing compound over plug threads, then install the plug in a sealing fitting to make it appear the seal has been poured.

Some new sealing compounds, particularly cable sealing materials, use epoxy. The manufacturer's instructions should be followed.

Explosionproof enclosures often use special high-strength bolts. Substitution of an equivalent size bolt of a different strength can reduce the effectiveness of the enclosure. The bolt strength is usually indicated by a coded marking on the head of the bolt. If a bolt is lost or damaged, be sure it is replaced by a bolt of the same size and strength.

Pipe-taping materials should not be applied to threaded explosionproof joints. If corrosion protection is desired, or if there is a need to make the threaded joint more resistant to moisture penetration, a corrosion-inhibiting grease (such as petrolatum or soap-thickened mineral oil) may be applied to the joint surfaces before assembly. The grease should be of a type that does not harden because of aging, contain an evaporating solvent, or cause corrosion of the joint surfaces. These same materials can also be used to inhibit corrosion of other types of joints. Manufacturers may provide joints with other types of corrosion-protecting materials that have been tested as part of the explosionproof enclosure.

Some lines of explosionproof equipment are made so that parts are interchangeable, and in other lines parts are not interchangeable. Care should be exercised to be sure that all parts of a particular piece of explosionproof equipment are designed for that particular piece of equipment and have been tested as part of that equipment. For example, covers for one box may fit another, but they may not have been tested for that other box. Explosionproof attachment plugs should be used only with the receptacles for which they have been tested, as indicated by the marking on the products.

Most important, of course, is to be sure that the equipment is identified as being suitable for the particular hazardous location involved (class and group) and that it is not limited to Division 2 locations only. Equipment intended for Division 1 locations may or may not be identified with the division of the hazardous location. Equipment suitable only for Class II locations is not suitable for use in Class I locations.

After installation, routine maintenance procedures are still needed, particularly because many hazardous locations are also corrosive locations. The enclosure should be checked for signs of unusual corrosion and replaced if necessary. Joint surfaces, because they often have a lower degree of corrosion resistance than the enclosure itself, are particularly subject to corrosion and should be checked and cleaned if necessary. Threaded joints, particularly in conduit systems, should be checked visually.

Grounding continuity should be checked visually (and if necessary, by tests) wherever possible. If tests are conducted to determine the adequacy of the grounding connection, care should be taken that there are no flammables present outside or inside the system during the testing procedure.

Bolts and nuts should be checked to see that they are all in place and properly tightened. All enclosures should be checked to make sure that they are complete, that covers are not missing, plugs not left out, and so on.

Other Considerations: Explosionproof equipment costs more to build than the equivalent nonhazardous location equipment, and this cost is passed on to the purchaser. It is desirable to design an installation so that a minimum of explosionproof equipment is needed. This can be done by careful attention to area classification and location of equipment and by using nonelectrical equipment or other protection systems that may be less costly. The larger the equipment, the less likely it is that explosionproof equipment will be available.

There is a great variety of explosionproof equipment available for use in Group D locations, because this is the most common Class I hazardous location. There is also a considerable amount of equipment, other than motors and generators, suitable for Group C locations. Motors and generators suitable for Group C locations are quite limited in availability.

There are no motors and generators of the explosionproof type listed for Groups B or A locations. This does not mean they cannot be designed. The design problems can be, and in some special circumstances have been, overcome. The problem is that the very close tolerances for the shaft joints cannot usually be maintained on a production-line basis. Also, because the demand for Group B or Group A motors is extremely limited, there is no incentive for the manufacturers to design motors suitable for these locations.

4-2.1.2 Intrinsically Safe Equipment and Wiring

Definition: The *National Electrical Code* defines intrinsically safe circuits as circuits in which any spark or thermal effect is incapable of causing ignition of a mixture of flammable or combustible material in air under prescribed test conditions.

ANSI/UL 913[5], the standard used to investigate intrinsic safety, separately defines intrinsically safe circuits, intrinsically safe apparatus, normal operation, and fault conditions.

Intrinsically safe circuits are defined as circuits in which any spark or thermal effect, produced either normally or in specified fault conditions, is incapable, under the test conditions prescribed in the standard, of causing ignition of a mixture of flammable or combustible material in air in the mixture's most easily ignited concentration. Intrinsically safe apparatus is defined as apparatus in which all circuits are intrinsically safe.

Associated apparatus is that apparatus in which the circuits are not necessarily intrinsically safe themselves, but that affect the energy in the intrinsically safe circuits and are relied upon to maintain intrinsic safety. A typical example of associated apparatus is an intrinsic safety barrier, which is a network designed to limit the energy available to the protected circuit in the hazardous location under specified fault conditions. (See Figures 4-42 and 4-43 later in this section, under Intrinsically Safe Barriers.)

Associated electrical apparatus may be either of the following:

1. Electrical apparatus that has an alternative type of protection for use in the appropriate potentially flammable atmosphere
2. Electrical apparatus not so protected that is not permitted to be used within a potentially flammable atmosphere

Normal operation is defined as intrinsically safe apparatus or associated apparatus conforming electrically and mechanically with its design specification. Opening, shorting, or grounding of the intrinsically safe wiring itself is considered a normal condition.

A fault is defined as a defect or electrical breakdown of any component, spacing, or insulation that alone or in combination with other faults may adversely affect the electrical or thermal characteristics of the intrinsically safe circuit. If a defect or breakdown leads to defects or breakdowns in other components, the primary and subsequent defects and breakdowns are considered to be a single fault.

System Concept: Intrinsically safe equipment and wiring is a system concept. Any piece of equipment connected to the system or that can be connected to the system is part of the system. The only exception is self-contained battery-operated equipment that is not connected to any other equipment or apparatus.

A thermocouple, even though it may generate only millivolts at extremely low current levels, is not intrinsically safe in itself, because it is normally connected to a voltmeter or potentiometer, which may, in turn, be connected to other circuits. If the chart drive-motor wiring of a recording potentiometer, which often operates at 120 V, should somehow contact the thermocouple leads within the potentiometer enclosure, the energy from the 120-V circuit would be available on the thermocouple, which could be energized to 120 V above ground as a result of the fault. Fault conditions within associated apparatus, such as the recording potentiometer, are considered when evaluating intrinsically safe systems. Even though the thermocouple itself may not be capable of developing sufficient energy to cause ignition of a flammable atmosphere, such ignition-capable energy may be available as a result of connection of the thermocouple to other equipment when a fault occurs within the equipment to which the thermocouple is connected.

As indicated in Section 2-8, a very small but measurable amount of energy is needed to cause ignition of a flammable atmosphere. The concept of the intrinsically safe protection system is to maintain the available energy below that required to cause ignition, even under fault conditions.

Except for high-voltage circuits (those operating at several thousand volts), where the spark between energized parts or an energized part and a grounded object may result in very little absorption of energy by the arcing parts, there is usually considerable energy absorbed by these arcing parts, or "electrodes." Also, the energy is highly unlikely to be released instantaneously in a normal arcing situation. If a spark-testing apparatus could be designed to represent the worst-case sparking condition likely to occur, and if the circuit parameters, that is, voltage and energy storage (capacitance or inductance) capability, were defined, the minimum ignition energy of any particular flammable mixture could be redefined as the minimum igniting current under a specified set of conditions. Such test equipment has been designed and is used in the United States and internationally for testing intrinsically safe apparatus, as explained below. The apparatus is used to determine whether or not the energy available in a particular circuit is ignition-capable.

In addition to arcs or sparks as a result of shorting, opening, or grounding of circuits, high surface temperatures can be a source of ignition. Even with low-energy equipment, temperatures well above the ignition temperature of flammable gases and vapors can be attained by heating small-diameter wires or wire strands under short-circuit conditions, or as a result of thermal runaway of solid-state components, such as transistors. One only has to consider the extremely high temperature attained by a flashlight bulb filament to recognize this potential ignition source.

Simply defining a voltage or a current level is an insufficient definition of intrinsic safety, because intrinsic safety is an energy concept. Thus, time and energy storage are brought into the picture. Even a 1.5-V carbon-zinc dry-cell battery is capable of delivering sufficient energy under certain conditions to cause ignition of a flammable atmosphere. The reader may recall simple experiments conducted in science classes in which students make electromagnets by winding insulated wire around an iron nail and connecting the wire to a battery. When the battery circuit is interrupted, a visible spark at the opening "contacts" appears as a result of release of the energy stored in the (inductive) electromagnet.

Battery-operated flashlights and lanterns, even those using small dry-cell batteries, are not inherently intrinsically safe. Fine wire strands may short-circuit the battery if the design is not proper, and the combination of the temperature and spark when the wire burns open may result in ignition. A short-circuit, or even normal current in some of the larger lanterns, may produce sufficient energy at switch contacts to cause ignition, depending upon the flashlight design and the amount of energy absorbed at the contacts.

Possibly the most severe condition is breakage of the flashlight or lantern bulb, which operates at incandescent temperatures. Tests have shown that if the lamp bulb is broken in a flammable atmosphere without simultaneously breaking the lamp filament, the combination of high temperature and the spark created when the filament burns open as a result of exposure to the oxygen in the air can result in ignition, even of Group D materials. This problem can be overcome in two ways: by incorporating a spring mechanism that will disconnect the lamp bulb from the battery if the bulb glass is broken or by designing the lamp compartment so that breakage of the lamp bulb is highly unlikely. Listed flashlights and lanterns for use in hazardous locations employ such protection techniques.

Conditions Considered: When considering the acceptability of intrinsically safe equipment and circuits and associated apparatus, both normal and abnormal conditions of operation are considered. Sources of spark ignition are considered from discharge of capacitive circuits, interruption of inductive circuits, intermittent making and breaking of resistive circuits, and hot-wire fusing. The sources of thermal ignition considered are heating of small-gauge wires strands, glowing of a filament, and high surface temperature of components.

Normal operation is considered to include all of the following:

1. Supply voltage at maximum rated value
2. Environmental conditions within the ratings given for the intrinsically safe apparatus or associated apparatus
3. Tolerances of all components in the combination that represents the most unfavorable condition
4. Adjustments in the most unfavorable settings
5. Opening of any one of the field wires, shorting of any two field wires, and grounding of any one of the field wires of the intrinsically safe circuit being evaluated

Abnormal operation includes one or two faults introduced into the equipment, such as shorting primary to secondary of a transformer, open-circuiting of a shunt diode, short-circuiting of a capacitor or resistor, or short-circuiting or grounding any part of the circuit. However, protective components and sufficient spacing or equivalent between energized parts can be considered as not being subject to fault. For example, if there is adequate creepage and clearance distance between live parts on a printed circuit board, that distance can be considered as not subject to fault.

Transformers with certain construction and meeting certain performance requirements can be considered not subject to primary-to-secondary fault. So, too, can current-limiting resistors and blocking capacitors that are subject to specific tests to determine that they are not likely to fail in an unsafe manner. Because only a maximum of two separate independent faults are considered, redundancy of components (for example, resistors in series or shunt diodes in parallel) is a common design technique.

Testing: A number of different types of tests are conducted on intrinsically safe circuits and apparatus, as well as associated apparatus. Dielectric withstand tests are conducted on blocking capacitors to determine whether or not they can be considered protective components and therefore not subject to faults. Several different types of tests are conducted on transformers to determine whether or not they can be considered protective components and not subject to faults. Shunt diode barriers are also subjected to a number of tests. The final and most critical test is the spark ignition test to determine whether or not, after consideration of protective components and the application of a variety of safety factors, the circuit is capable of ignition.

The test apparatus used for the spark ignition test is that which has been agreed upon internationally and is included in the International Electrotechnical Commission (IEC) requirements and in ANSI/UL 913. It consists of an explosion chamber of about 15.25 in.^{3} (250 cc) volume, in which circuit making-and-breaking sparks can be produced in the presence of a prescribed test gas representing the hazardous location group involved. Components of the contact arrangement are a cadmium disc with two slots and four tungsten wires 0.2 mm (0.008 in.) in diameter, which slide over the disc. (See Figures 4-32 and 4-33.) The free length of the tungsten wires is 11 mm (0.44 in.), 1 mm (0.039 in.) greater than the separation between the spindle driving the tungsten wires and the slotted disc. The driving spindle, to which the tungsten wires are attached, makes 80 revolutions per minute. The spindle on which the cadmium disc is mounted revolves in the opposite direction. The ratio of the speeds of the driving spindle to the disc spindle is 50 to 12, so that the disc operates at 19.2 revolutions per minute in a direction opposite to the rotation of the tungsten wires. The spindles are insulated from one another and from the housing. The explosion chamber is designed to withstand pressures up to 213 psi (1470 kPa) or provided with suitable pressure relief. If cadmium, zinc, or magnesium will not be present in the circuit under test, the cadmium disc is permitted to be replaced by a tin disc. When test currents are high, the tungsten wires tend to limit the current, so that other materials (such as copper) may be needed for the wire "whiskers."

The specific gas-air mixtures used in the spark ignition test are 5.25

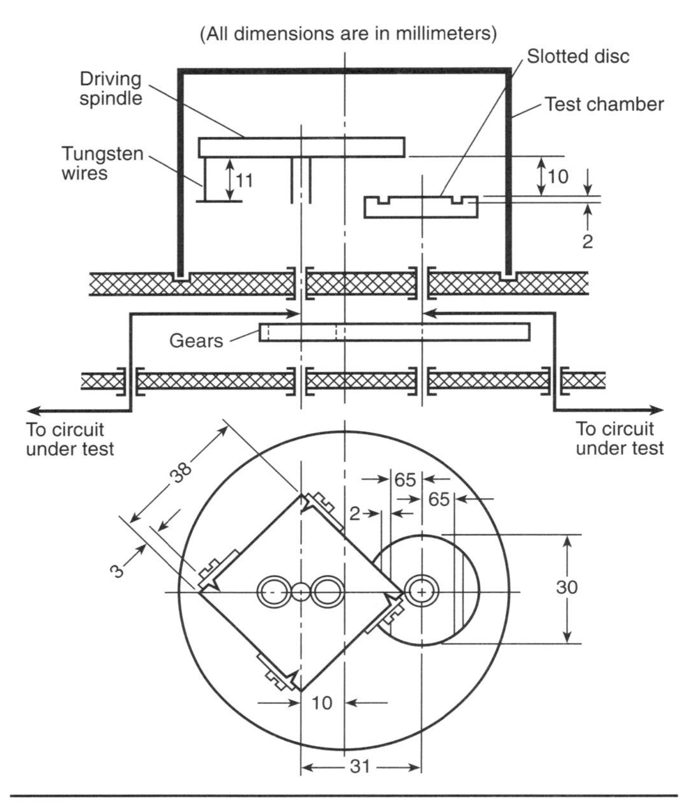

Figure 4-32. *A schematic of spark testing apparatus.*

±0.25 percent propane in air for Group D; 7.8 ±0.5 percent ethylene in air for Group C; and 21 ±2 percent hydrogen in air for Groups A and B. All percentages are on the basis of volume. Gases of at least 95 percent purity are required. If the equipment under test is intended for use only in a specific gas or vapor and is appropriately marked, it can be tested with the most easily ignited concentration of that gas or vapor in air. Oxygen can be substituted for or added to the air if the test is for apparatus for use in an oxygen-enriched atmosphere.

The sensitivity of the spark test apparatus is checked before and after each test series, using a 24-V dc circuit containing a 0.095 henry air-core coil. The currents in the circuits are set in accordance with Table 4-2.

The particular circuit and equipment used, as well as the adjustments in current, are based on the equipment and circuit that is to be tested. For the calibration, the spark test apparatus is run for 400 revolutions

Figure 4-33. *Spark testing apparatus.*

(5 minutes) of the tungsten wire holder with the holder at positive polarity.

Once the apparatus has been calibrated, it is inserted in the circuit to be tested. For example, if consideration is to be given to whether or not a circuit is ignition-capable with reference to grounded parts, the

Table 4-2 Circuit Currents for Spark Test Apparatus

	Must Ignite		Must Not Ignite	
Group	Cadmium Disc	Tin Disc	Cadmium Disc	Tin Disc
D	100 mA	110 mA	71 mA	98 mA
C	65 mA	90 mA	49 mA	77 mA
A and B	30 mA	50 mA	25.5 mA	41 mA

apparatus is inserted in the circuit with one terminal (either the tungsten wire holder or the cadmium or tin disc) connected to the circuit and the other terminal connected to ground. For direct current circuits, the tungsten wire holder is run for not less than 400 revolutions (5 minutes), 200 at each polarity. For alternating current circuits, 1000 total revolutions (12.5 minutes of test) is required. During the testing, a safety factor of 1.5 is normally used. This safety factor may be achieved by different methods, such as decreasing the values of limiting resistances to obtain 1.5 times the circuit current, increasing the line voltage by 1.5, and so forth. No ignitions are permitted during actual spark ignition testing.

Since there is considerable data available on minimum igniting currents for various circuits using the standardized test apparatus, a comparison to this data is permitted so that actual spark testing is not necessary on some circuits. This information can be used by equipment designers to provide reasonable assurance that the equipment either will pass the test or will be judged as not requiring tests because the energy available is sufficiently below the ignition energy, making testing unnecessary.

The assessment as to whether or not the data, in the form of ignition curves, can be used to eliminate actual spark testing is limited to those circuits that can be readily assessed in terms of the elementary circuits represented by the curves. The circuit conditions for assessment include all normal and fault conditions, except the additional safety factor of 1.5 mentioned above. In its place, a different safety factor is used. For normal conditions and single-fault operation, the current (or voltage, if applicable) is not permitted to exceed a specified percentage of the value determined from the appropriate curve. For two fault conditions, the current is not permitted to exceed a specified (but higher) percentage of the value determined from the appropriate curve. The curves and currents used for this evaluation are shown in Figures 4-34 through 4-41. All voltages and currents are dc or peak ac.

In addition to determining whether or not a circuit is ignition-capable, either by comparison to the appropriate curves or actual spark testing, and in addition to the various tests to determine whether or not

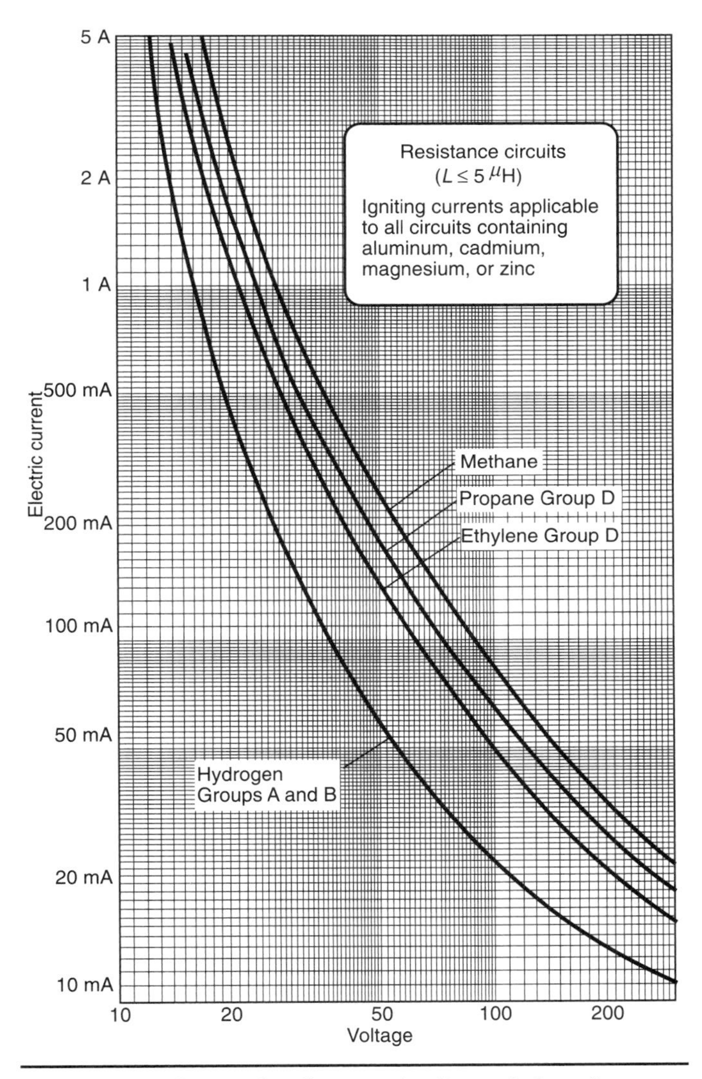

Figure 4-34. *Curve of resistance circuits, cadmium disc.*

a particular component can be considered a reliable component and therefore not subject to fault during considerations of possible faults, temperature tests may be required on components likely to exceed the ignition temperature of the gas or vapor involved. For very small components, where the component temperature may exceed the known ignition temperature of the gas or vapor involved, actual ignition testing is permitted. Small heated surfaces can exceed the ignition temperature

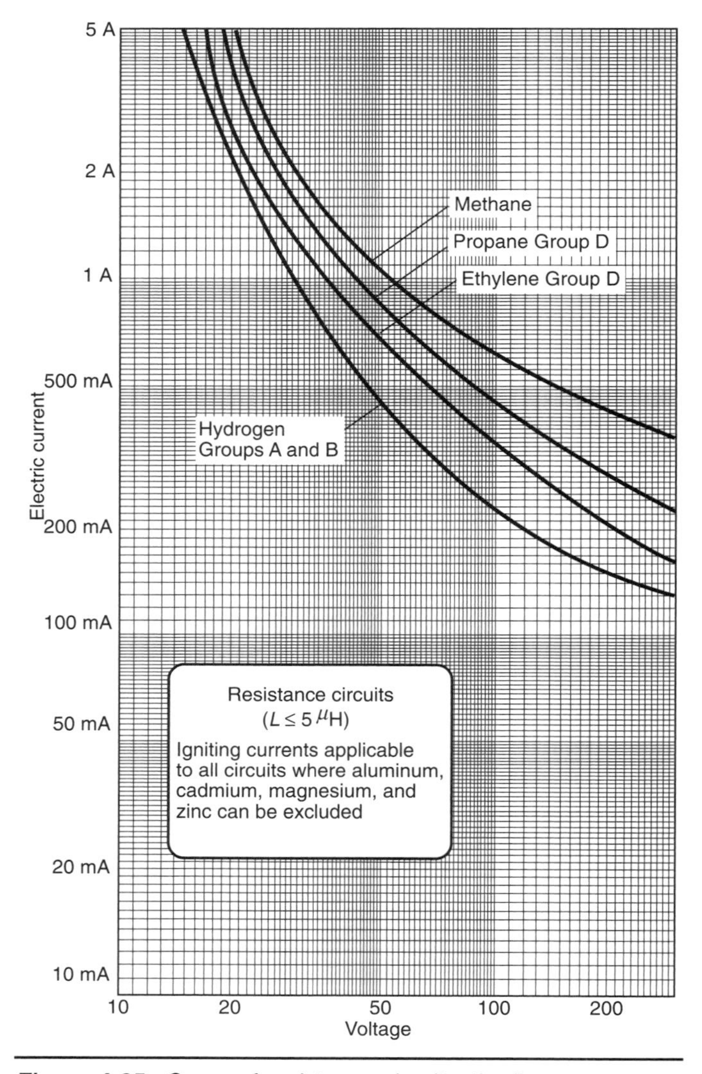

Figure 4-35. *Curve of resistance circuits, tin disc.*

without causing ignition because of turbulence at the heated surface and the method by which the ignition temperature of a flammable material is determined. (See Chapter 2.)

Dielectric withstand tests, mechanical tests of partitions, and battery ejection drop tests (for portable handheld, battery-operated apparatus of certain constructions) are also conducted.

The most difficult part of the investigative procedure is usually

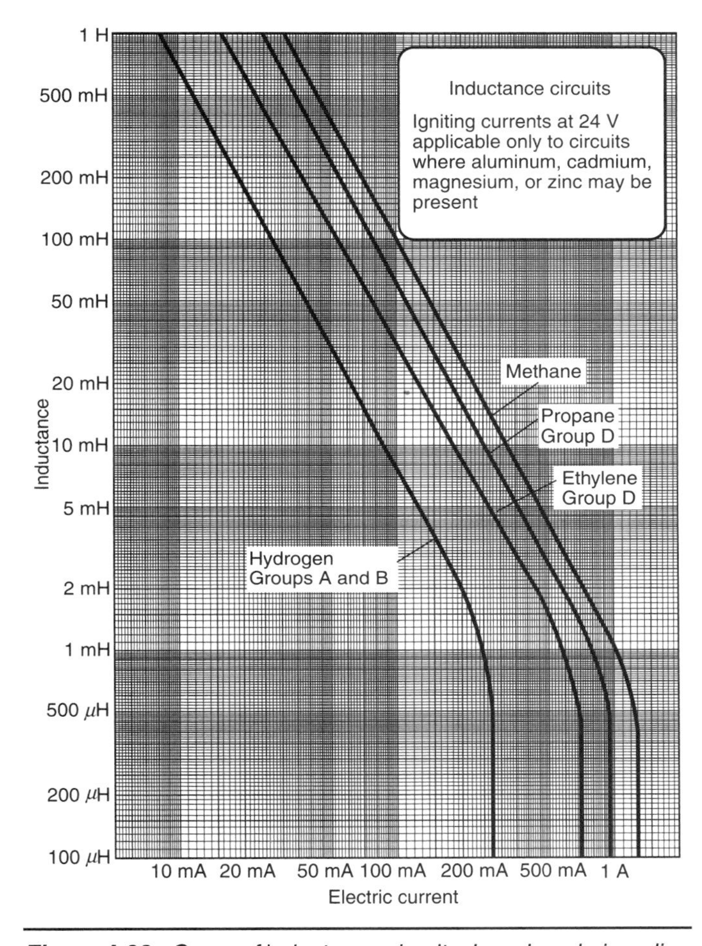

Figure 4-36. *Curve of inductance circuits, L vs. I, cadmium disc.*

determining worst-condition faults in complex electronic apparatus. This may require a complete fault analysis of the equipment. The fault analysis often can be simplified by making certain assumptions. For example, all capacitors in a circuit can be assumed to be in parallel through some type of single fault, and the circuit can be evaluated with a simple capacitor with a value equal to the sum of all capacitors. The same can be done with series inductors. With the extremely low levels of energy needed to operate modern solid-state electronic equipment, these assumptions usually can be made without resulting in a test failure. They greatly simplify the evaluation task.

Installation: The *National Electrical Code* indicates that equipment and associated wiring approved as intrinsically safe is permitted in any

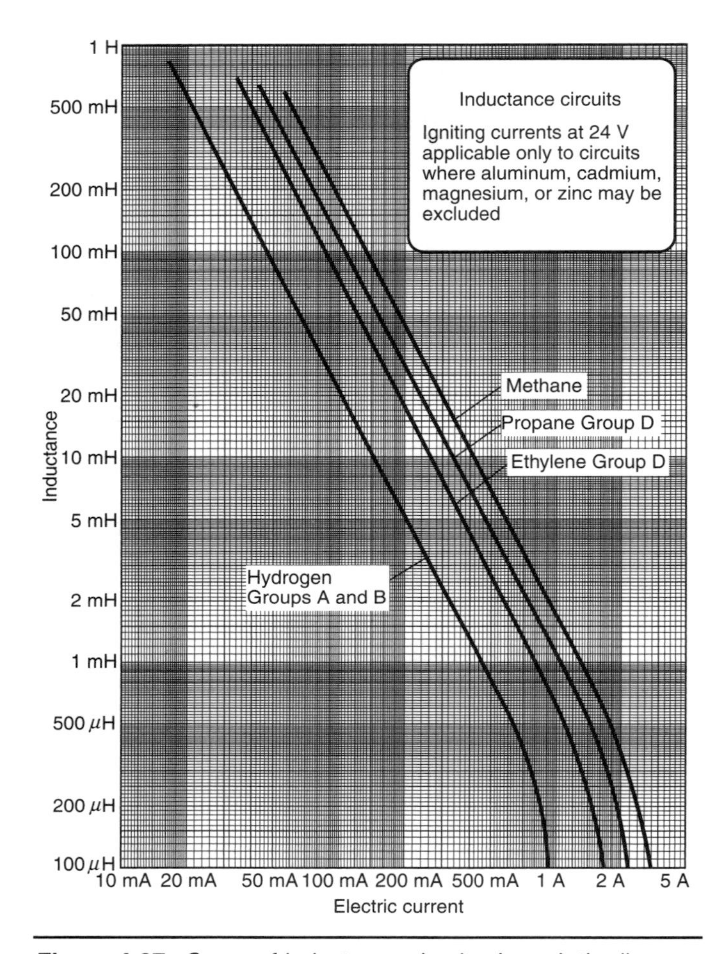

Figure 4-37. *Curve of inductance circuits, L vs. I, tin disc.*

hazardous location for which it is approved, and the provisions of Articles 500, 501, 502, and 503 are not considered applicable to such installations. In addition, the *NEC* requires that wiring of intrinsically safe circuits be physically separated from wiring of all other circuits that are not intrinsically safe and that means be provided to prevent the passage of gases and vapors.

The key to installation of intrinsically safe circuits is to take precautions to prevent intrusion of unsafe energy from other circuits. These other circuits may be different intrinsically safe circuits (intrinsically safe circuits in which the possible interconnections have not been evaluated and approved as intrinsically safe) or nonintrinsically safe circuits. Intrusion of unsafe energy is more likely to occur in a nonhazardous

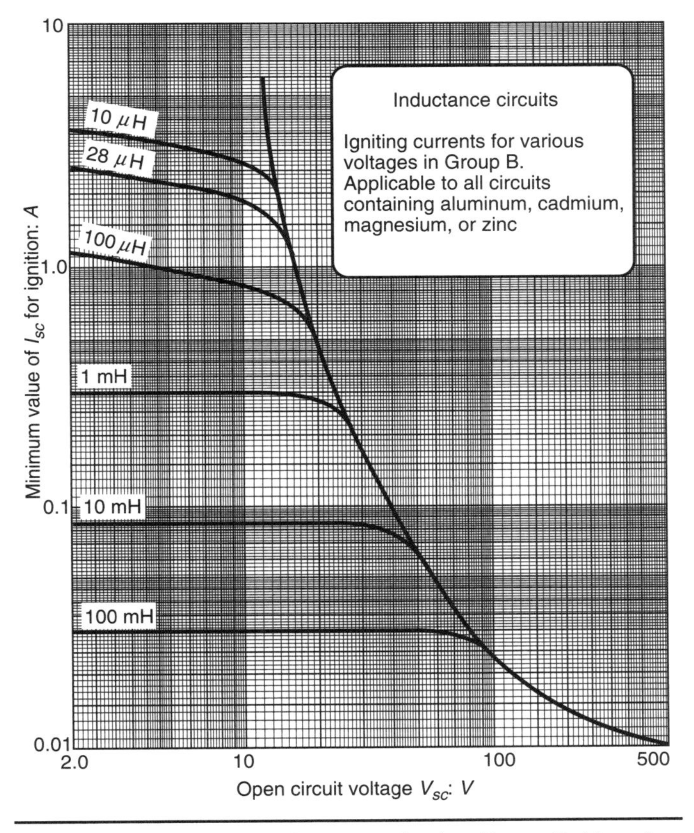

Figure 4-38. *Curve of inductance circuits, Group B, V vs. I, cadmium disc.*

location than it is in a hazardous location, because the wiring methods for nonintrinsically safe circuits in hazardous locations are more restrictive than they are in nonhazardous locations. For example, ordinary (nonintrinsically safe) Class 2 signal or control circuits can be run as open wiring in nonhazardous locations, yet they can easily provide more than sufficient energy to cause ignition in a hazardous location should the nonintrinsically safe Class 2 wiring and intrinsically safe wiring come together without adequate insulation. Therefore, ANSI/ISA RP12.6[6] specifies that in nonhazardous locations intrinsically safe and nonintrinsically safe wiring external to panels be in separate enclosures, cables, raceways, or cable trays. The enclosures, cables, raceways, or cable trays for intrinsically safe wiring are to be identified as containing intrinsically safe wiring. The intrinsically safe and nonintrinsically safe wiring may occupy the same enclosure, raceway, or cable tray if they

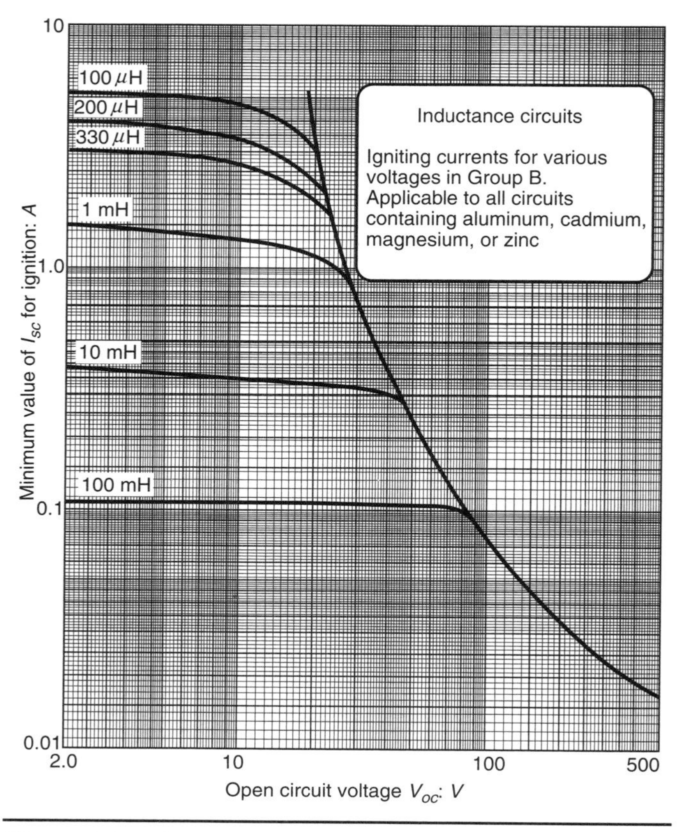

Figure 4-39. *Curve of inductance circuits, methane, V vs. I, cadmium disc.*

are permanently spaced at least 50 mm (2 in.) apart and tied down separately. Similar separation requirements, although not quite as restrictive, are applicable to intrinsically safe wiring in panels, such as control panels.

The *National Electrical Code* also requires that, where intrinsically safe wiring is employed, means be provided to prevent transmission of gases or vapors. This is based on the sealing requirements of the *NEC*, which are designed to prevent a cable or raceway from transmitting a flammable gas or vapor from a hazardous to a less hazardous or nonhazardous location through the raceway or cable. For sealing cables and raceways containing intrinsically safe circuits, ordinary conduit fittings for sealing can be used, as can other means, such as ventilated enclosures or nonexplosionproof seals.

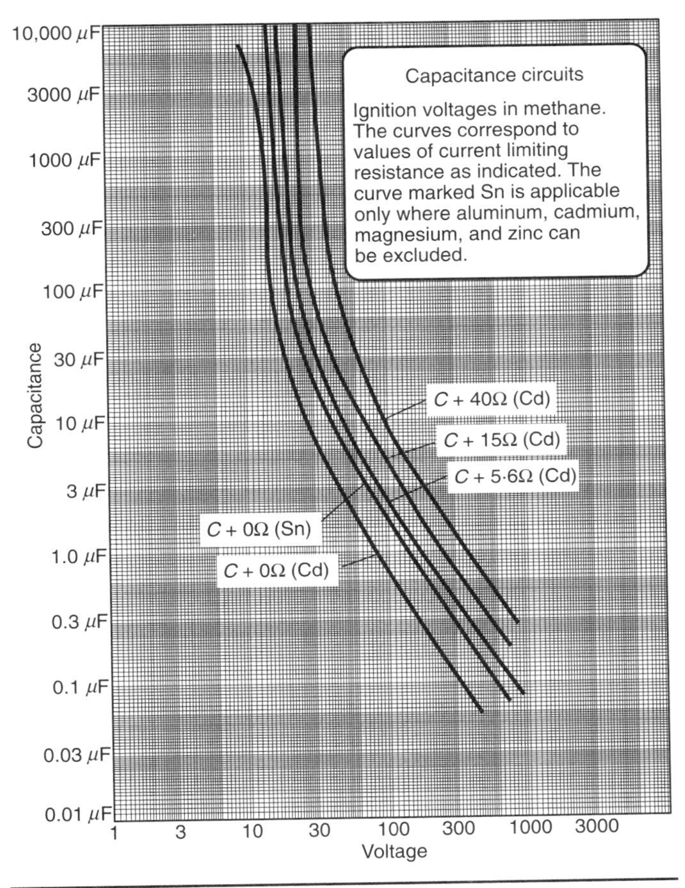

Figure 4-40. *Curve of capacitance circuits, V vs. C, methane, tin disc.*

Intrinsically Safe Barriers: An intrinsically safe barrier system, the most common of which is a shunt diode barrier, is a common and convenient method of establishing intrinsically safe circuits. It permits use of a variety of equipment on the nonhazardous location side of the barrier without the need for a laboratory-type evaluation of all parts on this side of the system. It also permits a variety of equipment to be installed on the hazardous location side of the barrier, provided energy storage characteristics of this equipment are known.

The most common shunt diode barrier system is the zener diode barrier. This may be either fused or unfused, as shown in Figures 4-42 and 4-43. F_1 in Figure 4-43 and R_0 in Figure 4-42 limit the current in Z_1 and Z_2 to the power rating of the zener diodes under fault conditions

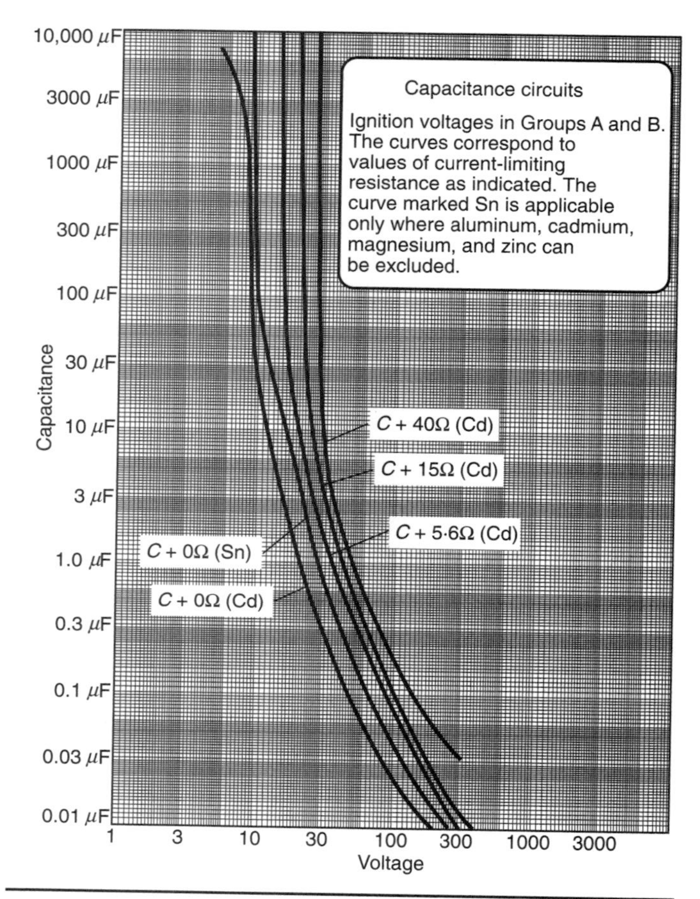

Figure 4-41. *Curve of capacitance circuits Groups A and B, V vs. C, tin disc.*

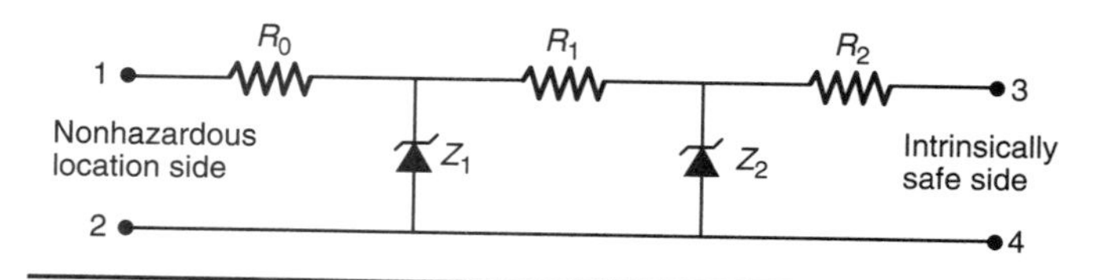

Figure 4-42. *Unfused zener diode barrier.*

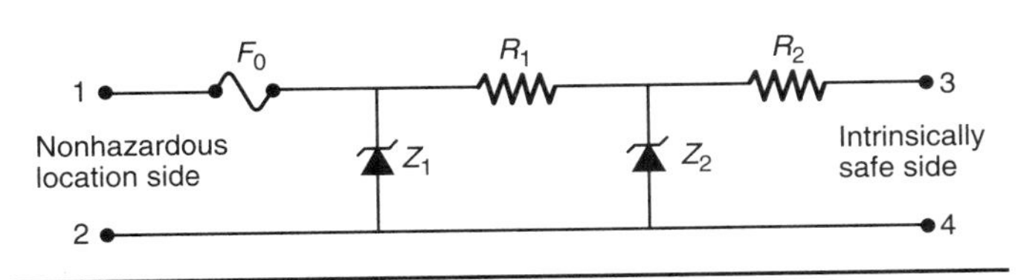

Figure 4-43. *Fused zener diode barrier.*

so that the diodes need not be excessively large. The resistor R_0 is investigated as a reliable component not subject to short-circuit fault. Although the diodes themselves can be considered subject to open-circuit fault, they are redundant components, and either one alone will provide the necessary protection in the event of a first fault, that is, high voltage across input terminals 1 and 2. Resistor R_2 is the current-limiting resistor, and it, too, is investigated as a reliable component not subject to short-circuit fault. Resistor R_1 is used primarily to permit testing to determine that the diodes are working properly. Terminals 2 and 4 and the ends of the two diodes are connected to a ground bus, which in turn is connected to a signal ground system to which all grounds in the intrinsically safe system are connected. A very low impedance (1 ohm or less is usually recommended) is necessary so that the voltage level on the ground bus will not be raised to an unsafe level under high-current fault conditions.

A shunt diode barrier system is commonly tested with 250 V ac across terminals 1 and 2, representing a fault in auxiliary equipment on the nonhazardous location side of the barrier that imposes a voltage up to 250 V ac on these terminals. Diode Z_1 conducts at its rated voltage, typically 14 or 15 V for a barrier designed for a 12-V (normal operation) system, thus limiting the output voltage of the circuit to 14 or 15 V. The voltage at which this diode conducts is designed to be higher than the rated input voltage (under no fault conditions) at terminals 1 and 2. The fuse in a fused barrier is selected so that it will open before the power rating of Z_1 is exceeded. Diode Z_2 usually conducts within one or two volts of Z_1, and it serves as a backup to Z_1 in the event Z_1 fails in an open-circuit condition for any reason.

Resistor R_2 limits the current in the intrinsically safe circuit to the required value at the output voltage of Z_1 and Z_2. Thus, even though up to 250 V ac may be applied to the input of the circuit (terminals 1 and 2) as a result of a fault in the equipment or circuit connected to terminals 1 and 2, the output at terminals 3 and 4 cannot exceed the voltage and current permitted by the diodes and the resistor.

Shunt diode barriers normally have maximum inductance and capacitance ratings on the intrinsically safe side, because even though both the voltage and the current are limited at terminals 3 and 4, too much inductance in the circuit connected to terminals 3 and 4 could result in release of an ignition-capable spark when the circuit opens. In a similar manner, too much capacitance between terminals 3 and 4 could result in release of an ignition-capable spark if there was a short-circuit between wiring connected to terminals 3 and 4 or between wiring connected to terminal 3 and ground.

Wiring always has inductance and capacitance associated with it, depending upon the spacing between conductors, size of conductors, length of conductors, and so on . Therefore, it is necessary to limit the length of conductors connected to terminals 3 and 4, just as it is necessary to limit the inductance and capacitance of connected equipment. The length limitation is usually in the order of thousands of feet. Manufacturers instructions provide the information on installation limitations.

By adjusting the values of the barrier's components, it can be designed for a variety of uses, including different types of instrument systems and thermocouples. Such barriers are available commercially from a number of different manufacturers.

The barrier system permits a variety of equipment to be connected to terminals 3 and 4, provided that the capacitance and inductance of this equipment and interconnecting wiring is within the barrier rating. The barrier system also permits a variety of equipment to be connected to the system supplying terminals 1 and 2, provided that the maximum voltage possible in the equipment connected to terminals 1 and 2 does not exceed the barrier maximum input voltage (usually 250 V ac) under any condition of fault. Usually, this maximum voltage can be determined easily by knowledge of the voltage of the branch circuit to which the equipment is connected. However, if the apparatus connected to terminals 1 and 2 includes cathode-ray tubes or other equipment with voltages exceeding 250 V ac generated within the equipment, additional precautions are necessary and ordinary barriers alone may not be suitable.

Only two very simple types of barriers are described above. There are others, some of them active electronic barriers instead of the passive barriers described. Optical barriers are becoming more common. Power supplies and other equipment designed to supply intrinsically safe wiring often include barriers or partial barriers within the equipment, simplifying the equipment design and reducing the time and cost of investigating the equipment.

Maintenance Precautions: Intrinsically safe systems need maintenance to ensure continued safety. The word "intrinsically" in "instrinsically safe" is something of a misnomer, because it implies that no matter what is done, the circuit is safe. This is not entirely true. Only two faults are considered when investigating intrinsically safe equipment, other than opening, shorting, or grounding of field-installed wiring. Therefore, if there are three or more separate and independent faults within the intrinsically safe or associated equipment, safety may be compromised. In Figure 4-42, if the circuits to both Z_1 and Z_2 or the diodes themselves both open, and there is then a third fault that imposes high voltage on the system, the high voltage will appear across terminals 3 and 4. Resistor R_2 is designed to limit the current to a safe value at the conducting voltage of Z_1 or Z_2, not at a higher voltage.

The first priority in maintaining an intrinsically safe system is to check to see that all safety components of the system are intact, including both diodes in a zener diode barrier. This is the purpose of resistor R_1, as shown in Figures 4-42 and 4-43.

The grounding system is also essential in maintaining intrinsic safety, and it should be checked. Caution must be taken when checking with instruments that are connected to or provided with energy systems (such as batteries) to assure that the testing itself does not compromise safety and does not result in damage to any part of the system.

Terminals and connections should be checked to see that they are tight and that there are no stray strands of wire that could short-circuit the terminals. Separation between intrinsically safe and nonintrinsically safe circuits, both in hazardous and in nonhazardous locations, should be checked to be sure that there have been no changes from the original design (such as nonintrinsically safe wiring put into a cable tray with intrinsically safe wiring).

4-2.1.3 Purged and Pressurized Equipment

Definition: The *National Electrical Code* does not include a specific definition of purged and pressurized equipment. However, it is included as a recognized protection technique in the *NEC*, which also indicates that in some cases, hazards may be reduced or hazardous locations limited or eliminated by adequate positive-pressure ventilation from a source of clean air in conjunction with effective safeguards against ventilation failure. The *NEC* refers to NFPA 496[7], *Standard for Purged and Pressurized Enclosures for Electrical Equipment,* for further information.

The *NEC* also indicates that nonexplosionproof motors and generators may be used in Class I, Division 1 locations if of the totally enclosed type supplied with positive pressure ventilation from a source of clean air with discharge to a safe area, so arranged as to prevent energizing the machine until ventilation has been established and the enclosure has been purged with at least 10 volumes of air, and also arranged to automatically deenergize the equipment when the air supply fails. This is an excellent description of one type of purged equipment.

In addition, the *NEC* also permits a nonexplosionproof totally enclosed inert gas-filled type of motor or generator supplied with a suitable reliable source of inert gas for pressurizing the enclosure, with devices provided to ensure a positive pressure in the enclosure, and arranged to automatically deenergize the equipment when the gas supply fails. This, too, is an excellent description of a type of purged equipment.

Therefore, even though the *NEC* does not specifically define purged and pressurized equipment, it includes descriptions of protection methods that are, in effect, definitions of the two different methods of protection by purging recognized in NFPA 496.

In the first method, the equipment or enclosure is purged by a continuous flow of clean air or an inert gas until flammables have been eliminated from the enclosure. The flow is then continued at sufficient volume to maintain a pressure of at least 25 Pa (0.0036 psi or 0.1 in. of water) inside the enclosure. This method is particularly suitable for equipment that requires ventilation to remove heat, such as motors, generators, and transformers, and for control rooms where people work within the enclosures.

The second method of protection is essentially the same, except that instead of maintaining a continuous flow after the initial purge, the enclosure is kept pressurized to at least 25 Pa (0.0036 psi or 0.1 in. of water) with a sufficient supply of pressurized clean air or inert gas to maintain the pressure in the event of small leaks in the enclosure. This method of protection is more suitable to small enclosures, such as those for instruments.

The protection method with continuous flow can be used either where the flammable gas or vapor may surround the enclosure or where the gas or vapor source is from within the enclosure, such as may occur with gas analyzers.

NFPA 496 recognizes two different levels of protection in Class I, Division 1 locations: Type X purging and Type Y purging. (See Figure 4-44.) Type X purging changes the classification of the interior of the enclosure from Division 1 to nonhazardous. Type Y purging changes the classification of the interior of the enclosure from Division 1 to Division 2. The differences in requirements are the following:

1. In Type X purging, there must be a pressure or flow detecting device that, upon sensing inadequate flow or pressure, will automatically disconnect all electric power to the enclosure.
2. In Type Y purging, the flow or pressure sensing device is not required to automatically disconnect all electrical supply sources within the enclosure, but it is required to initiate an audible or visible alarm indicating loss of flow or pressure. Since Type Y purging changes the interior classification from Division 1 to Division 2, all equipment inside the enclosure must be suitable for Division 2 locations.

The flow- or pressure-sensing device, if it is electrical in nature, will be in a Division 1 location during the initial purge operation and when

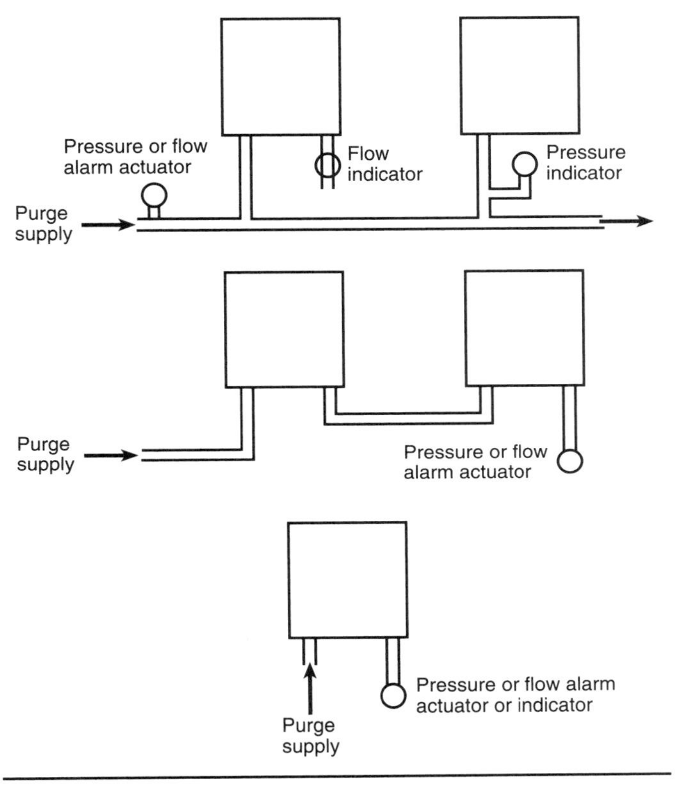

Figure 4-44. *Typical installations for Type Y purging.*

there is a loss of flow or pressure. This device is therefore required to be suitable for the Division 1 location.

The gas used for purging and maintaining pressure may be either a source of clean air or an inert gas. For small enclosures, particularly instrumentation to be used in corrosive atmospheres, dry nitrogen is a common gas used where a positive pressure is maintained without flow other than the initial purge. Clean air is the most common gas used where a continuous flow is maintained, although flue gases have been used for purging the tanks of ships at sea.

Although automatic disconnection of all electrical power to equipment that is not suitable for Division 1 locations in the purged and pressurized enclosure is required in Type X purging, it is still possible for components operating at high temperatures to cause ignition if they are immediately exposed to a surrounding flammable vapor after disconnection from the electric supply source. Warnings on the equipment are required where such components are housed to indicate that the enclosure is not to be opened until the parts have cooled below the ignition temperature. Similar precautions are necessary for Type Y purged and pressurized enclosures.

Purging and pressurizing of enclosures to make them suitable for use in Class I, Division 1 hazardous locations is basically a field installation technique rather than a matter of purchasing equipment specifically manufactured to be suitable for use in Class I, Division 1 locations. A general exception is gas analyzers and other equipment where the source of the flammable gas or vapor is within rather than around the outside of the equipment. Also, the pressure or flow sensing devices and disconnect or alarm actuators are normally ''off-the-shelf'' explosionproof devices or devices intended to be used in intrinsically safe circuits, if the devices are electrical in nature.

All enclosures for purged and pressurized equipment are required to be of noncombustible material with sufficient strength to withstand the abuses likely to be encountered. The enclosures must also be strong enough to withstand the internal pressures, and precautions must be taken to protect the enclosure from excessive pressure of the purged supply. Windows should be of tempered glass or other shatterproof material.

Installation and Maintenance: During and immediately after installation, the purged and pressurized equipment should be checked to be sure that it operates as intended and that automatic timers intended to delay reenergizing the equipment after loss of pressure are properly set. If the enclosure is divided into separate compartments, the flow should be checked to ensure that the purging operation purges any possible flammable gases from the separate compartments.

Proper maintenance includes checking the intake system for corrosion, clogged lines, and leaks that might introduce flammable gases or vapors into the enclosure, as well as the usual maintenance of motors, driving fans, and compressors.

4-2.1.4 Other Methods of Protection

There are other methods of protection for equipment for use in Class I, Division 1 locations that are not, strictly speaking, explosionproof, intrinsically safe, or purged and pressurized systems. Such protection systems are usually limited to specific items of equipment that have been thoroughly investigated and determined acceptable by qualified experts, with adequate production control of the system at the factory.

One such system is the common hermetic motor compressor used for refrigeration and air-conditioning equipment. The electrical components outside of the hermetically sealed motor-compressor unit are protected by conventional protection systems, usually explosionproof enclosures.

The inside is considered adequately protected because of the overload and overtemperature systems provided for the motor, the extensive production control and manufacturing precautions, and factory testing of such equipment to maintain the hermetic seal.

Electrostatic paint-spraying equipment is not intrinsically safe as defined in ANSI/UL 913, yet such equipment has been investigated and listed or approved by qualified electrical testing laboratories for many years. The laboratories conduct extensive tests, and these, together with careful control of the installation, have resulted in adequate levels of safety.

Attachment plugs and receptacles are another example of equipment that is not, strictly speaking, completely explosionproof; however, extensive testing and control of the design has indicated that they are safe for use in Class I, Division 1 locations. Some receptacles depend upon automatic disconnecting switches in explosionproof enclosures, so that when the plug is inserted or removed from the receptacle, the arc that results from making and breaking the load will not be at the plug- and receptacle-contacting surfaces. Others are designed so that the arc at the contacts occurs deep within the mated parts, with long flame paths and tight clearances — in effect resulting in an explosionproof enclosure during insertion and withdrawal and when the two parts are mated. However, the receptacle contacts themselves may not be totally enclosed in an explosionproof enclosure when there is no plug in the receptacle. Also, the point where the flexible cord enters the attachment plug is not "explosionproof" as defined in the *NEC*.

Another type of protection for nonexplosionproof motors in Class I, Division 1 locations is recognized in the *National Electrical Code*. This protection is for motors designed to be submerged either in a liquid that is flammable only when vaporized and mixed with air or in a gas or vapor at a pressure greater than atmospheric and that is flammable only when mixed with air. The machine is required to be arranged to prevent it from being energized until it has been purged with the flammable liquid or gas to exclude air, and the arrangement also automatically deenergizes the equipment when the supply of liquid or gas or vapor fails or the pressure is reduced to atmospheric. This protection system is designed specifically for motors immersed in, for example, liquefied or pressurized natural gas. It is not intended for use where the liquid is water. When properly designed and maintained, ignition is prevented by maintaining the flammable mixture above the upper flammable limit. Precautions are necessary during those periods when the tank is being filled or emptied and the mixture within the tank may be within the flammable range.

There are also flashlights and lanterns that are neither explosionproof nor intrinsically safe. They are described in Section 4-2.1.2.

4-2.2 Division 2

Because a Class I, Division 2 location is one where ignitible concentrations of flammable gases or vapors exist only under an abnormal condition, the equipment requirements and wiring methods are not as restrictive as in Class I, Division 1 locations.

4-2.2.1 Equipment Suitable for Division 1 Locations

Equipment suitable for Class I, Division 1 locations is also suitable for use in Class I, Division 2 locations of the same group. It also may be suitable for use in other groups, but if so used should be treated as general-purpose or nonhazardous location equipment. For example, in a Class I, Group C, Division 2 location, an ordinary open or enclosed motor (not specifically identified as suitable for use in hazardous locations) may be used if it does not contain switches, overload or overtemperature devices, slip rings, or commutators, all of which could produce arcs or sparks under normal conditions. Since a Class I, Group D explosionproof motor may be the equivalent of such a general-purpose motor, it may be used in a Class I, Group C, Division 2 location. However, it must be treated just as an ordinary location motor is treated; that is, there should be no commutators, slip rings, and so on, that are arcing or sparking devices. Even though such arcing or sparking devices would be enclosed in a Group D explosionproof enclosure, a Group D explosionproof enclosure is not suitable for use in a Group C location.

The *National Electrical Code* requires that if equipment in a Class I, Division 2 location is an ignition source under normal conditions of equipment operation, it must be approved for Class I, Division 1 locations, with certain exceptions as noted in Sections 4-2.2.2 through 4-2.2.4.

4-2.2.2 Nonincendive Circuits

The word ''nonincendive'' is found in very few dictionaries. It means not capable of causing ignition under normal conditions of operation. A nonincendive circuit is defined in the *NEC* as a circuit in which any arc or thermal effect produced, under intended operating conditions of the equipment or due to opening, shorting, or grounding of field wiring, is not capable, under specified test conditions, of igniting the flammable gas, vapor, or dust-air mixture.

The concept is the same as with other equipment for Class I, Division 2 locations: a fault in the equipment that makes it capable of causing ignition is very unlikely to occur at the same time as a fault that produces a flammable atmosphere in a Division 2 location.

Nonincendive circuits are sometimes referred to as intrinsically safe circuits for Division 2 locations only, even though ''intrinsically safe'' by definition includes consideration of fault conditions.

In addition to nonincendive circuits as defined above, there are nonincendive contacts, nonincendive equipment, and nonincendive components. (See Section 4-2.2.3.)

A nonincendive circuit is, because of its definition, a low-energy circuit. If it were not, opening, shorting, or grounding one or more of the circuit conductors could result in the release of ignition-capable energy. Nonincendive contacts are not necessary in a nonincendive circuit, but they are necessary if there are contacts in a circuit that is a low-energy circuit. The circuit itself may be capable of releasing ignition-capable energy when tested using a spark testing device. Nonincendive contacts are contacts large enough to absorb some of the energy that would otherwise be released into the atmosphere, and, when tested, the contacts do not release ignition-capable energy into the atmosphere surrounding the contacts.

An example of nonincendive contacts that may be in a circuit that is not nonincendive is an ordinary nonelectronic (rotary dial) telephone connected to a central station. Although the energy available in the circuit, particularly under ringing conditions, may be capable of igniting a Group D atmosphere when using a spark testing apparatus such as shown in Figure 4-32, the various switching contacts in the telephone itself are not capable of releasing sufficient energy at the contacts to ignite a Group D atmosphere.

Nonincendive equipment is usually self-contained, battery-operated, or solar cell-operated equipment that, although it may not meet all of the requirements for intrinsic safety, will meet the requirements for nonincendive equipment, because only normal conditions of operation are considered. Most common types of handheld calculators, battery-operated watches, hearing aids, paging receivers, and even low-powered radio transceivers are examples. Such equipment is often intrinsically safe as well as nonincendive, but it may not have been evaluated as intrinsically safe.

Nonincendive equipment differs from so-called ''nonsparking'' equipment (see Section 4-2.2.8) in that nonsparking equipment includes splices in junction boxes and even squirrel-cage motors that do not have any arcing or sparking parts not otherwise protected. Nonsparking equipment is usually installed in ignition-capable circuits, and if the equipment should fault, it could cause ignition.

Care must be exercised in using nonincendive equipment in Class I, Division 2 locations. Such locations are usually adjacent to or above Division 1 locations. Dropping nonincendive portable equipment may result in a fault, releasing ignition-capable energy at the same time the equipment falls into a Division 1 location (a tank or pit, for example).

Nonincendive circuits and equipment are tested or evaluated essentially in the same way that intrinsically safe circuits and equipment are tested or evaluated, except that abnormal conditions are not considered. The spark testing apparatus is the same. Nonincendive contacts on the other hand, are tested either by actually operating the contacts in an explosive atmosphere or by substituting a suitable spark testing apparatus for the contact (This is only for contacts where it is believed that the circuit in which they are located is also nonincendive.)

The standards used in the United States to investigate equipment for Class I, Division 2 hazardous locations (except electric lighting fixtures) are ANSI/ISA S12.12[8] and UL 1604[9]. For electric lighting fixtures, the standard is ANSI/UL 844[10].

4-2.2.3 Nonincendive Components

The *NEC* defines a nonincendive component in the following manner:

> a component having contacts for making or breaking an incendive circuit and the contacting mechanism shall be constructed so that the component is incapable of igniting the specified flammable gas- or vapor-air mixture. The housing of a nonincendive component is not intended to (1) exclude the flammable mixture or (2) contain the explosion.

A typical nonincendive component is a micro-type switch. The very small molded plastic housing surrounding the switching mechanism does not have sufficient free volume for a flammable mixture inside the housing to be ignited, even if an ignition-capable spark is generated at the contacts. The size of the contacts may also absorb enough energy to prevent release of an ignition-capable spark.

Nonincendive components are tested by trying to create an explosion by arcing at the switching contacts when the component is in a flammable atmosphere, with a flammable mixture introduced into the component housing. The performance requirements are in UL 1604.

4-2.2.4 Oil Immersion

Immersion of ignition-capable contacts in oil is another method recognized by the *National Electrical Code* for protection of contacts that would otherwise be ignition-capable in Class I, Division 2 locations. ANSI/UL 698[11] provides construction and performance details.

4-2.2.5 Hermetically Sealed

Enclosing contacts in hermetically sealed chambers is yet another method for protection in Division 2 locations. For this use, hermetically sealed is defined as involving a fusion joint (welding, soldering, brazing, or fusion of glass to metal) rather than a gasketed joint. The reason for this is twofold. First, there is usually no reliable way of determining whether or not a gasket seal is still effective, particularly after exposure to solvent vapors. Second, gasketed enclosures are normally designed to be opened, and there is usually no effective way of determining that the gasket is still in place or that the product has been reassembled to ensure a continued hermetic seal.

The most common examples of this method of protection are mercury-tube switches and reed relays.

4-2.2.6 Purged and Pressurized Enclosures

NFPA 496 permits purged and pressurized enclosures to be used in Class I, Division 2 locations. The designation for enclosures used in Division 2 locations to reduce the atmosphere inside the enclosure to nonhazardous is Type Z purging. The requirements are essentially the same as those for Type Y purging, as noted in Section 4-2.1.3.

4-2.2.7 Fixtures

Electric lighting fixtures for use in Class I, Division 2 hazardous locations are treated somewhat differently from other equipment for use in Class I, Division 2 locations, primarily because most lighting fixtures are sources of heat, and it may be very difficult for a specifier, installer, or electrical inspector to determine whether or not the temperature of the lamp exceeds the ignition temperature of the particular flammable atmosphere involved. This was not quite as difficult a number of years ago when only fluorescent, incandescent, and a very few mercury vapor fixtures were involved.

Modern lighting involves a variety of illumination sources in addition to the ordinary long-tube-type fluorescent, incandescent, and mercury vapor fixtures. A wide variety of metal halide (high-intensity discharge) lighting, high-intensity incandescent lamps (quartz-iodine, halogen, and so on), and bent-tube-type flourescent lamps with a built-in ballast intended for use in Edison-base lampholders are available today. With this wide range of lighting sources and fixture designs, it became impractical to rely on "best guesses" as to surface temperatures of parts that might be exposed to flammable atmospheres. Also, changes in inspection procedures in industrial occupancies in the early 1970s in the United States as a result of the Occupational Safety and Health Act forced greater reliance on third-party testing and certification. The *National Electrical Code* still does not require that electric lighting fixtures in Class I, Division 2 locations be specifically approved for those locations, except for most portable lighting equipment (which is required to be suitable for Class I, Division 1 locations). However, the *NEC* recognizes that electric lighting fixtures for Class I, Division 2 locations should be treated differently by specifically indicating that such fixtures need not be marked to indicate the hazardous location group, but that they are required to be marked with the class, division, and operating temperature if over 100°C (212°F).

Electric lighting fixtures for use in Class I, Division 2 locations are essentially enclosed lighting fixtures with means provided for threaded conduit connection and with a marked operating temperature or temperature range (except for most fluorescent types), based on temperature tests. Temperatures are measured on the hottest part of the fixture, either inside or outside, but not inside hermetically sealed lamp bulbs, tubes, and so forth. Except for the marking, and the testing that leads up to the marking, such fixtures are similar to many totally enclosed industrial-type outdoor lighting fixtures that are and have been used for years in nonhazardous locations.

4-2.2.8 Ordinary Location Equipment

If the product is not an electric lighting fixture or a type of equipment that produces ignition-capable arcs or sparks under normal equipment operating conditions, ordinary-location (general-purpose) equipment is permitted by the *National Electrical Code* to be used in Class I, Division 2 hazardous locations. Such equipment is often referred to as "nonsparking" equipment. Open squirrel-cage motors or totally enclosed motors, fan-cooled (TEFC) or not, are permitted in Class I, Division 2 locations, provided they do not contain any sliding contacts; centrifugal or other types of switching mechanisms, including motor overcurrent, overload, and overtemperature devices; or integral resistance devices, either while starting or while running, unless the internal ignition sources are suitable for use in Class I, Division 2 locations. For example, an open motor with a commutator could be used if the commutator was enclosed in a suitable purged and pressurized enclosure or explosionproof enclosure. A low-surface-temperature heater in the motor, externally controlled, may be acceptable if the heater is suitable for use in Class I, Division 2 locations.

Nonhazardous-location outlet boxes, transformers, coils, and a wide variety of solid-state electronic control devices are suitable for use in Class I, Division 2 locations, without any special protection mechanisms and without any special marking or specific listing for use in hazardous locations.

4-2.2.9 Seals

The *National Electrical Code* requires seals in Class I, Division 2 locations for essentially the same reasons as they are required in Class I, Division 1 locations. (See Section 4-2.1.1.) However, the requirements for seals in Division 2 locations are somewhat more complex, because in addition to conduit wiring systems, special Type MC cable, and

Type MI cable, other cable wiring systems are permitted. Consideration therefore must be given to cable not in conduit as well as to cable in conduit and to the various types of cables to be used.

Cables acceptable for use in Class I, Division 2 locations in accordance with the *National Electrical Code* include cables capable of transmitting gases or vapors through the core of the cable and those not so capable. They also include cables with jackets that are gas/vaportight and those that are not.

When the rules for cable seals first appeared in the *National Electrical Code*, their intent was to treat the sealing of cables in a manner similar to the sealing of individual insulated conductors in conduit. When sealing individual conductors in conduit, the sealing compound is poured around the conductor insulation. Because there are spaces between each strand of a stranded conductor, called the interstices of the conductor strands (see Figure 4-45), a conduit seal does not act as

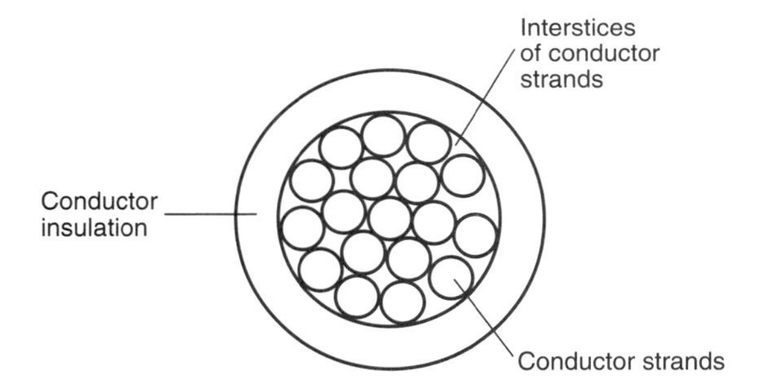

Single-insulated conductor cross section

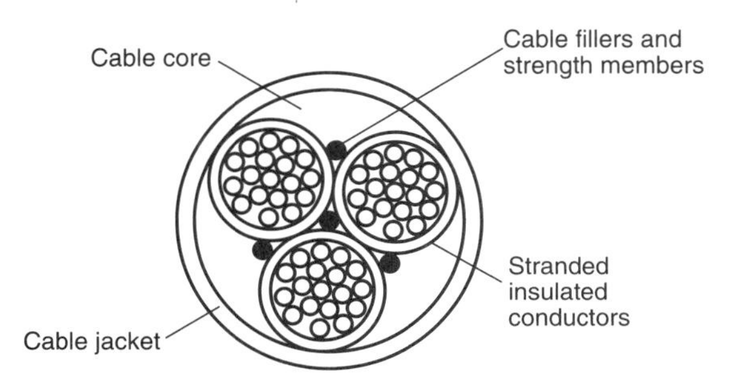

Multiconductor cable cross section

Figure 4-45. *When applying the requirements for cable seals, a multiconductor cable is treated in a similar manner as single conductors in conduit, except that recognition is given to cables that cannot transmit gases or vapors through the cable core.*

a complete block for gases and vapors. It reduces the passage of gases and vapors to a manageable level, provided the end of the conductor is not pressurized so as to force the gas or vapor through the interstices of the strands.

In the larger conductor sizes (those larger than No. 2 AWG), sufficient space in the interstices of the conductor strands may permit passage of an appreciable amount of gas — and even of weak explosions — if ordinary stranded conductors (that is, not compact stranded conductors) are used. This should be given consideration in critical situations. The strands can be sealed, for example, by using a compression-type connector and tape between the end of the connector and the end of the conductor insulation.

Multiconductor cables can be considered similar to insulated conductors in a conduit. In addition to the space between the individual strands of an insulated conductor, there may be space between the individual insulated conductors of a cable inside the outer jacket. This space is known as the cable core. (See Figure 4-45.)

The cable core may be closed up by other elements of the cable, such as fillers to make the cable round, and strength members, or it may be relatively open. The cable may, therefore, either be capable or incapable of transmitting gases or vapors through the cable core. The criteria as to whether or not the cable is capable of transmitting gases or vapors through its core is the cable performance as compared to a conduit seal for individual insulated conductors.

One of the tests for a conduit seal is determining that it will not pass 0.007 ft^3 of air per hour (198 cm^3 of air per hour) at a pressure of 6 in. of water [0.2166 psi (1493 Pa)]. Therefore, if the cable is constructed in such a manner, or the installation (such as the number of bends and clamps, vertical or horizontal run, and so on) is such that the core of the cable is equivalent to a conduit seal, the cable is considered a gas-blocking cable not capable of passing gases or vapors through its core. Some cables are specifically designed and listed as gas-blocking cables.

In addition, consideration is given to whether or not the jacket of the cable will prevent gases or vapors from entering the cable core. If a seal is used, but gases or vapors can enter the core of a cable capable of transmitting the gases or vapors through the cable jacket, the seal will not minimize the transmittal of the gases or vapors to a nonhazardous location as required by the *NEC*, unless it is located at the boundary of the Division 2 and nonhazardous location.

Even cables that have a core capable of transmitting gases or vapors in excess of the amount permitted for conduit seals are unlikely to transmit gases and vapors through the core, unless there is a pressure differential between the end of the cable in the Division 2 location and the end of the cable in the nonhazardous location, so as to force the gases and vapors through the core. The *National Electrical Code* recognizes this by indicating that a cable with a gas/vaportight jacket that is capable of transmitting gases or vapors through the cable core is not required to be sealed unless either it requires a seal because the cable enters an explosionproof enclosure, or the enclosure in the hazardous location is pressurized in such a manner as to subject the cable to a pressure in excess of 6 in. of water [0.2166 psi (1493 Pa)]. In this respect, cables are not treated the same as insulated conductors in conduit.

Since explosionproof enclosures may be required in a Class I, Division 2 location because the enclosed apparatus is ignition-capable under normal operating conditions, the same considerations must be given to sealing these enclosures in Division 2 locations as is given in Division 1 locations; that is, the explosionproof enclosure must be completed by a seal just as it is in a Division 1 location. If a cable enters the explosionproof enclosure, and the cable is not in conduit, the jacket must be removed and the individual insulated conductors of the cable must be sealed. Fittings are available that are designed for sealing multiconductor cables entering explosionproof enclosures, without the need to use a conduit fitting for sealing and a conduit nipple. (See Figure 4-15). If the cable is in conduit, the rules for sealing cables in Division 1 locations apply. See Figures 4-46 through 4-51 for various seal arrangements.

With the exception of metal-clad (Type MC) cable that has an

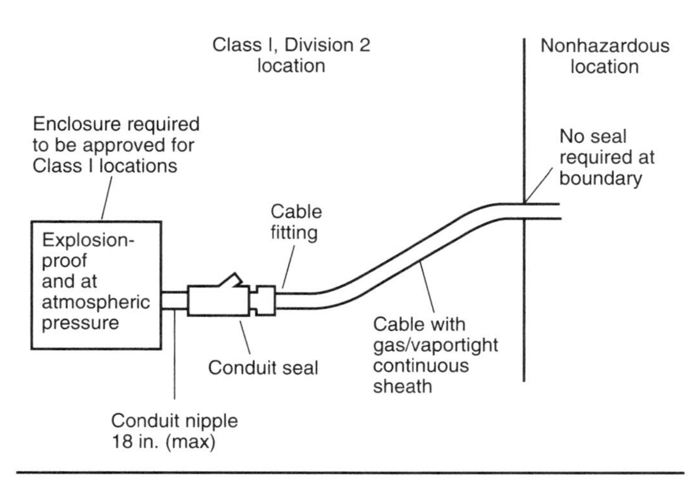

Figure 4-46. *In Class I, Division 2 locations, a seal at the boundary of the Division 2 and nonhazardous location is not required if the cable is sealed at the enclosure and has a continuous gas/vaportight sheath. The cable seal may be a cable sealing fitting for use in Division 2 hazardous locations instead of a combination of conduit nipple or equal, conduit seal, and cable fitting as shown in this drawing. (See Figure 4-47).*

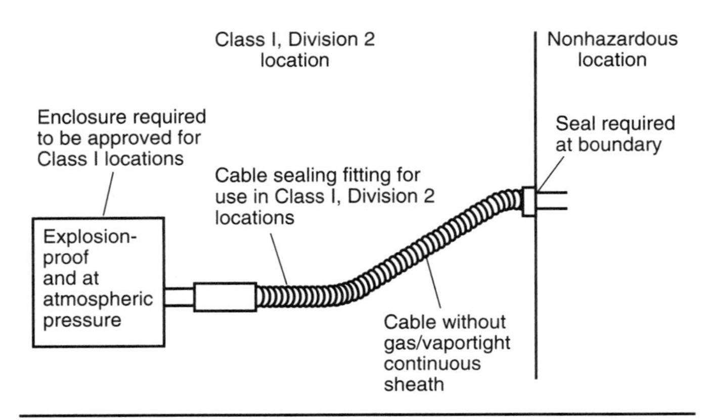

Figure 4-47. *If the cable does not have a gas/vaportight continuous sheath, for example, Type MC cable with interlocked armor, the seal is required at the boundary of the Division 2 and nonhazardous location.*

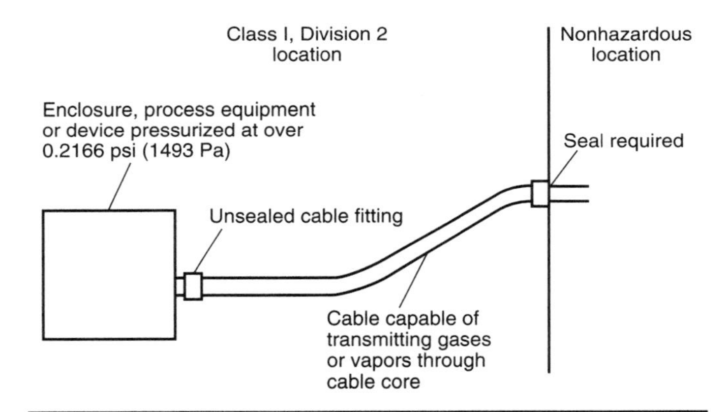

Figure 4-48. *If the cable is capable of transmitting gases or vapors through the cable core, and the end of the cable in the Division 2 location is pressurized [over 0.2166 psi (1493 Pa)], a seal is required between the enclosure and the boundary of the Division 2 location. If the cable does not have a gas/vaportight continuous sheath, the seal is required at the boundary; otherwise, it may be at any point between the cable fitting at the enclosure and the boundary.*

interlocked metal jacket and that is not otherwise protected by a continuous gas/vaportight sheath, all cables permitted for wiring in Class I, Division 2 locations are considered as having a gas/vaportight sheath. However, if the sheath is broken (opened) at any point and not sealed, the cable does not have a "*continuous* . . . sheath."

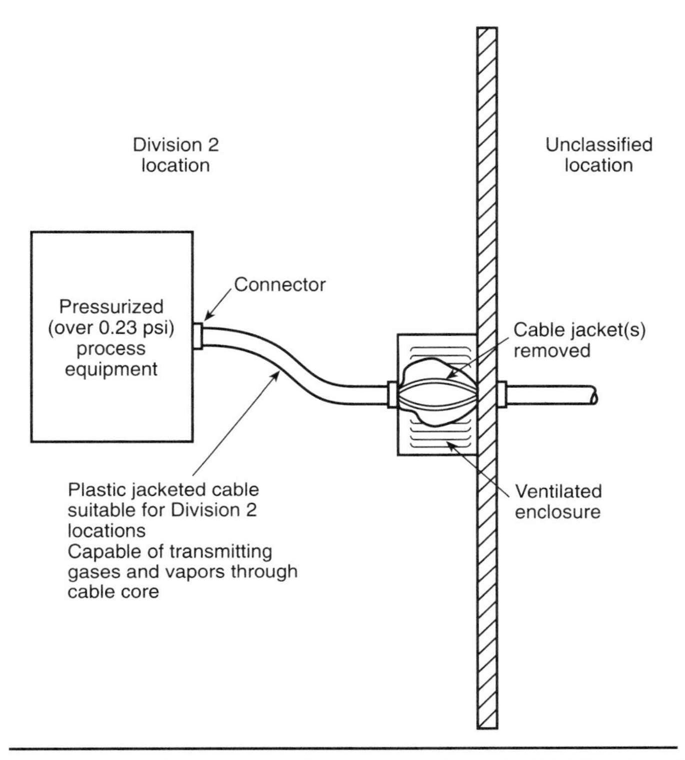

Figure 4-49. *The cable seal at the boundary of a Division 2 and nonhazardous location does not have to be an ordinary cable seal. "Barriers and other means" are permitted by the NEC, such as the ventilated enclosure shown. (Courtesy of HEP–Killark.)*

Cable seals, except for cables entering explosionproof enclosures, are not required to be explosionproof. Therefore, different methods of sealing can be used, provided the seal and sealing material used is suitable for the application. If a seal is required, for example, at the boundary of the Division 2 and nonhazardous location because the enclosure is pressurized and no other seal is provided, the seal at the boundary could be a ventilated enclosure as shown in Figure 4-49, arranged so that the pressure cannot be transmitted beyond the ventilated enclosure and so that any gases or vapors entering the core of the cable in the hazardous location will be unlikely to be transmitted into the nonhazardous location. Care must be exercised that transmittal of the gas or vapor through the interstices of the conductor strands will be minimized.

Seals in conduit and cable systems minimize, but do not prevent, the passage of gases and vapors through the conduit or cable. In addition to the interstices of the conductor strands, which are not sealed, and the

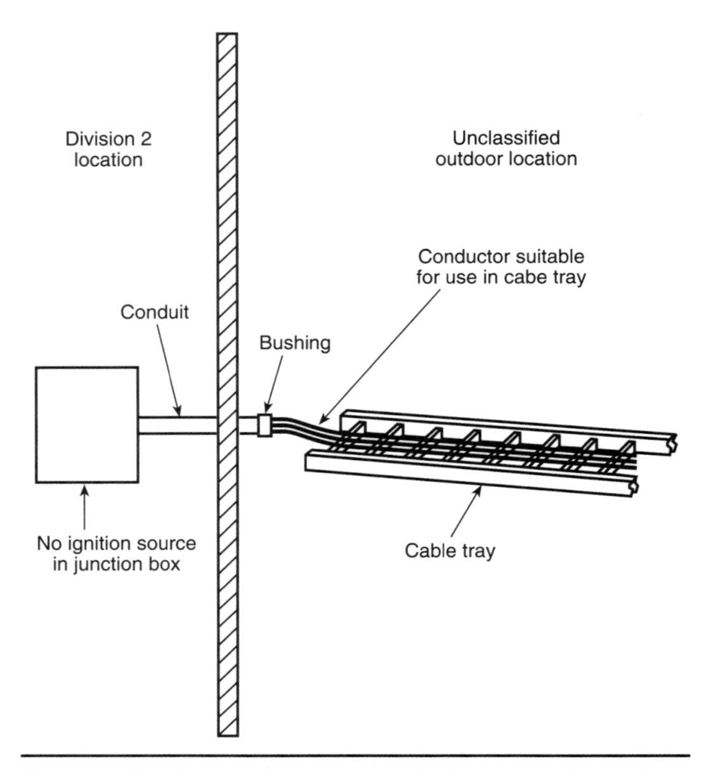

Figure 4-50. *No seal is required at the boundary for conduit wiring systems where a transition is made to a cable tray, cable bus, ventilated busway, Type MI cable, or open wiring in an unclassified outdoor location, or to an unclassified indoor location if the conduit system is all in one room. The conduit is not permitted to terminate at an enclosure that contains an ignition source under normal operating conditions. (Courtesy of HEP–Killark.)*

core of even a "gas-blocking" cable, which is simply a cable designed to prevent passage of more than 0.007 ft^3/per hour (198 cm^3/per hour) of air at a pressure of 6 in. of water [0.2166 psi (1493 Pa)], most sealing compounds are not designed to prevent gases and vapors from passing through them if there is a continuous pressure differential across the seal. Canned pumps and some process equipment are commonly pressurized, sometimes to many atmospheres. Wiring such equipment also commonly requires that electrical conductors enter the equipment. The equipment designer provides a primary seal or barrier to prevent the fluid from being forced out of the pressurized vessel through the wiring.

The *National Electrical Code* requirements indicate that in such circumstances, a secondary seal or barrier is required to prevent the flammable fluid from entering the electrical conduit system. It also

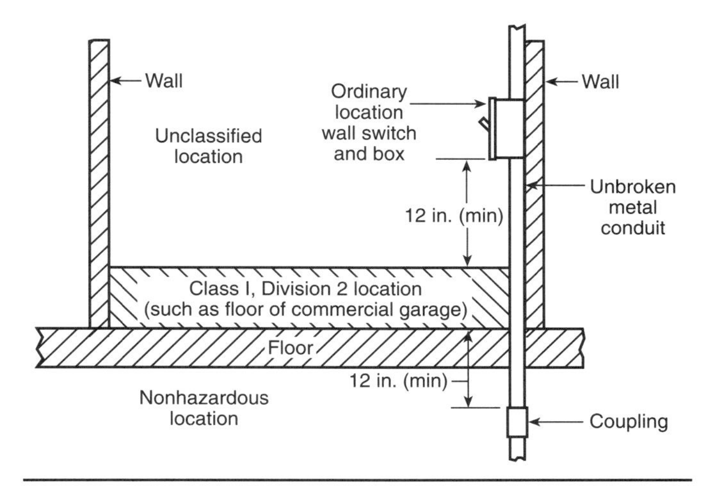

Figure 4-51. *If an unbroken metal conduit passes completely through a Division 2 location, seals are not necessary. (Courtesy of HEP–Killark.)*

requires a drain, vent, or other device so that primary seal leakage will be obvious. Ordinary conduit sealing fitting compound is somewhat porous, and it cannot serve to prevent or even particularly minimize the passage of gases and vapors if the pressure differential across the seal is appreciably greater than 6 in. of water [0.2166 psi (1493 Pa)]. Some other means must therefore be used to prevent passage of enough gases and vapors into a nonhazardous location from a Division 2 location, or from a Division 1 into a less hazardous or nonhazardous location, to present an explosion hazard. Figure 4-52 shows one method of accomplishing this.

4-2.2.10 Wiring Methods

In Class I, Division 2 locations, the wiring methods are less stringent than they are in Class I, Division 1 locations. Wiring methods suitable for Class I, Division 1 locations are acceptable. So, too, are enclosed and gasketed busways and wireways, cable tray systems, and a number of different types of cable. Where provision must be made for limited flexibility, as at motor terminals, flexible metal fittings suitable for Division 1 locations are permitted, as are flexible metal and nonmetallic conduit, liquidtight flexible metal conduit, and extra-hard service flexible cord. If the wiring is in a nonincendive circuit, nonhazardous location-type wiring methods are also permitted.

Ordinary (nonhazardous) location outlet and junction boxes and conduit and cable fittings are also permitted in Class I, Division 2 locations.

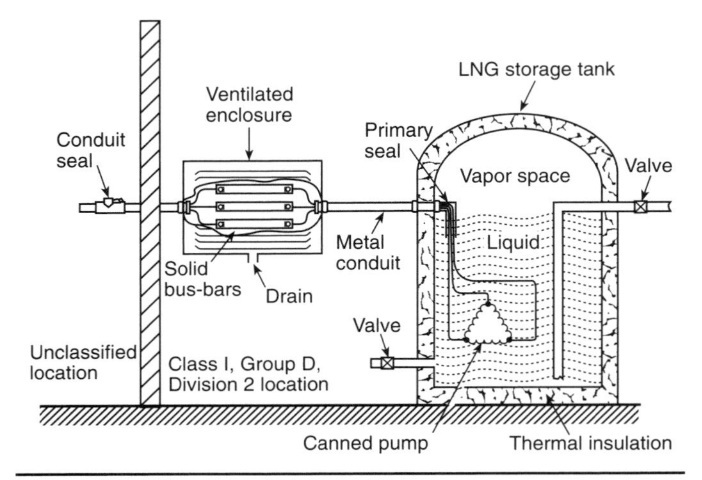

Figure 4-52. *A ventilated enclosure and solid busbars prevent the pressurized liquid in the canned pump from pressurizing the conduit system on the nonhazardous location side of the ventilated enclosure if the primary seal at the canned pump fails. They also prevent the flammable gas or liquid from the canned pump from being transmitted through the conduit seal. (Courtesy of HEP–Killark.)*

If flexible metal conduit or liquidtight flexible metal conduit is used, the conduit should not be depended upon as the equipment grounding path. Either the circuit should contain an equipment grounding conductor, or there should be a bonding conductor around the flexible metal conduit or liquidtight flexible metal conduit. The bonding conductor may be outside of the flexible section of raceway, but it should be securely fastened to the outside of the raceway and should be run physically in parallel to the axis of the raceway, not wrapped around the raceway. This will reduce to a minimum the inductance in the grounding circuit.

4-2.3 Zone 0

4-2.3.1 Wiring Systems

Very few wiring systems are recognized by Article 505 of the NEC for use in Zone 0 locations. Intrinsically safe equipment and wiring systems of the type of protection "ia" as described in Section 4-2.3.2 are acceptable. So, too, are optical fiber cables or systems with an approved energy-limited supply The reason for limiting the supply energy is to limit available light energy so that an ignition-capable "hot spot" of light cannot develop should the optical fiber cable be broken.

In addition, explosionproof wiring systems with all conductors in intrinsically safe or nonincendive circuits and with all conduits sealed (regardless of size) are currently permitted. However, consideration is being given to revisions of the *NEC* to eliminate these explosionproof wiring systems in Zone 0.

4-2.3.2 Intrinsic Safety

In Zone 0 locations, also suitable for Zone 1 and Zone 2, intrinsically safe equipment is almost identical to intrinsic safety protective systems as described in Section 4-2.1.2. This level is identified as type of protection "ia." Associated apparatus for type "ia" equipment is identified as Type "[ia]" apparatus.

The construction and performance requirements are in UL 913, IEC 79-11[12], and IEC 79-3[13].

4-2.4 Zone 1

4-2.4.1 Equipment and Wiring Suitable for Class I, Division 1

All equipment and wiring systems acceptable in Class I, Division 1 locations (see Section 4-2.1) and Class I, Zone 0 locations (see Section 4-2.3) are acceptable in Class I, Zone 1 locations. In addition, equipment specifically listed (as defined in the *NEC*) for Class I, Zone 1 is permitted. Such equipment is described in Sections 4-2.4.2 through 4-2.4.8.

4-2.4.2 Flameproof

Flameproof is defined as a type of protection of electrical equipment in which the enclosure will withstand an internal explosion of a flammable mixture that has penetrated into the interior, without suffering damage and without causing ignition, through any joints or structural openings in the enclosure, of an external explosive mixture consisting of one or more of the gases or vapors for which it is designed.

This protection method is quite similar to the explosionproof protection method (see Section 4-2.1.1), although the requirements differ somewhat and flameproof equipment is suitable only for Class I, Zone 1 and Zone 2 locations, Group IIA, IIB, or IIC, or a specific gas or vapor, depending on the construction and the gas or vapor used for testing.

The construction and test requirements are given in IEC 79-1[14], ISA S12.22.01[15], and UL 2279[16].

Flameproof enclosures are used primarily (although not exclusively) for protection where arcing and sparking contacts are present, such as on switches, circuit breakers, and motor starters. The enclosures are often high-quality reinforced plastic rather than the heavy cast metal normally used in explosionproof enclosures. The wiring compartments and parts other than arcing contacts are usually not in the flameproof enclosure itself. They are located in a separate enclosure protected by the increased-safety method of protection (see Section 4-2.4.3), with the conductors to the contacts passing through the flameproof enclosure wall via flameproof feed-through insulators.

The requirements for flameproof enclosures specify that the enclosure withstand 1.5 times maximum measured explosion pressure (with 100 percent testing at the factory), rather than four times maximum measured explosion pressure (on the samples submitted for proof testing at the testing laboratory). Fewer explosion tests are conducted, but they are performed with specified concentration of flammable mixtures. Also, it is assumed that cable wiring systems will be used so that there is no explosion testing with lengths of rigid conduit attached. Otherwise, the requirements are quite similar to requirements for explosionproof enclosures.

4-2.4.3 Increased Safety

Increased safety is defined as a type of protection applied to electrical equipment that does not produce arcs or sparks in normal service and under specified abnormal conditions, in which additional measures are applied so as to give increased security against the possibility of excessive temperatures and the occurence of arcs and sparks.

This protection method is different from any other previously described. It is suitable for use only in Class I, Zone 1 or Zone 2 locations. Since this protection technique is not sensitive to the explosion characteristics of the gas or vapor involved, except for the ignition temperature, it is suitable for use in Groups IIA, IIB, and IIC, and equipment using only this protection technique is not required to be marked with any Group II gas group. However, increased-safety equipment is often combined with other protection techniques, such as flameproof, so the gas group marking may still be needed (see Section 4-2.4.2).

The construction and test requirements are given in IEC 79-7[17], ISA S12.16.01[18], and UL 2279.

The increased safety method of protection is used widely for a variety of equipment, including, but not limited to, fluorescent lighting fixtures, motors and generators (with no arcing or sparking parts), junction and terminal boxes, transformers, solenoids, control equipment, and batteries. Some relatively low-temperature incandescent lighting fixtures can also be protected by the increased-safety technique.

The first of the additional measures applied to provide increased safety concerns the enclosure. Enclosures are required to meet certain impact tests and to be protected against harmful dust entry and water splashing from any direction.

Very careful attention is paid to field wiring terminals. Only clamping-type pressure wire connectors are permitted, and they must meet a variety of requirements to provide a high degree of protection against twisting, overheating, or loosening of connections. Internal (factory-installed) connections must also meet requirements designed to prevent loosening.

The characteristics of the connectors used for cable or conduit wiring, including the type, impact resistance, and protection against entry of water and dust, are specified in the installation instructions provided with the equipment.

Spacings (that is, creepage and clearance distances) between live parts and between live parts and grounded parts are required to be greater than in the equivalent equipment for ordinary locations, and high-grade insulating materials not subject to arc tracking are required.

For products that can be overloaded, such as motors, special protection against overheating is required, usually requiring installation only with the specific control equipment with which the increased safety motors have been tested under overload and stalled rotor conditions.

Special insulation and impregnation requirements are applied to transformer, coil, motor, and generator windings. Even the internal construction and the air gap between the rotor and stator of rotating electrical machinery is controlled in increased-safety equipment.

4-2.4.4 Intrinsic Safety

For Zone 1 locations, as well as for Zone 2, the level of intrinsic safety is identified as type of protection ''ib.'' It is the same type of protection as '' ia'' (see Section 4-2.1.2), except one less fault is introduced into the circuit before spark testing. The requirements are in IEC 79-11 and IEC 79-3.

Associated apparatus for Type ''ib'' equipment is identified as Type ''[ib]'' apparatus.

4-2.4.5 Purged and Pressurized

This protection technique is the same as for Class I, Division 1 locations as described in Section 4-2.1.3. It is also suitable for Zone 2 locations.

In the IEC system it is known as type of protection ''p.'' It is defined as a type of protection of electrical equipment that uses the technique of guarding against the ingress of the external atmosphere, which may be explosive, into an enclosure by maintaining a protective gas therein at a pressure above that of the external atmosphere.

The standards are IEC 79-2[19] and 79-13[20].

4-2.4.6 Oil Immersed

This protection technique is known as type of protection ''o.'' It is also suitable for Zone 2 locations. It is a type of protection in which the electrical equipment, or parts of the equipment, are immersed in a protective liquid in such a way that an explosive atmosphere that may be above the liquid or outside the enclosure cannot be ignited. It is similar to oil immersion as described in Section 4-2.2.4. A major use for this protection technique is for electrical contacts, particularly high-voltage contacts in high-current circuits.

The construction and test requirements are given in IEC 79-6[21], ISA S12.26.01[22], and UL 2279.

4-2.4.7 Powder Filling

Powder filling is identified as type of protection "q." It is a type of protection in which parts capable of igniting an explosive atmosphere are fixed in position and completely surrounded by filling material (glass or quartz powder) to prevent ignition of an explosive atmosphere. There is no equivalent system of protection recognized in *NEC* Article 500. It is used for protection of fuses and for filling junction boxes containing wire connections.

The construction and performance requirements are given in IEC 79-5[23], ISA S12.25.01[24], and UL 2279.

4-2.4.8 Encapsulation

Encapsulation, known as type of protection "m," is a protection system in which parts that could ignite an explosive atmosphere by either sparking or heating are enclosed in a compound in such a way that this explosive atmosphere cannot be ignited. There is no equivalent protection system recognized in *NEC* Article 500.

The construction and test requirements are given in IEC 79-18[25], ISA S12.23.01[26], and UL 2279.

4-2.5 Zone 2

This area classification corresponds to Class I, Division 2, and the same equipment and wiring systems are acceptable. (See Section 4-2.2.) So, too, are equipment and wiring suitable for Class I, Zones 0 and 1 locations. (See Sections 4-2.3 and 4-2.4.)

Equipment for Class I, Zone 2 locations only is identified as equipment with type of protection "n," with a suffix letter further identifying the type of protection, as follows.

4-2.5.1 Nonsparking Equipment

This type of protection is identified as type of protection "nA." It is the same type of equipment recognized in the *National Electrical Code* as suitable for use in Class I, Division 2 locations, because it does not have any normally arcing or sparking parts or parts operating at high temperatures.

4-2.5.2 Sparking Equipment

Equipment that contains normally sparking or arcing parts in which the contacts are suitably protected other than by restricted breathing enclosure are identified as type of protection "nC." This includes the nonincendive components as recognized in the *National Electrical Code* for Class I, Division 2 locations, and may include contacts protected by other means, such as those suitable for Class I, Division 1 or Zone 1 locations.

4-2.5.3 Restricted Breathing Enclosures

This type of protection is identified as type of protection "nR." It includes hermetically sealed enclosures as recognized in the *National Electrical Code* for Class I, Division 2 locations and other enclosures that pass tests and construction requirements designed to assure that flammable gases and vapors will not enter the enclosure, or they will enter so slowly that there is little likelihood of a hazard.

4-3 Class II Locations

4-3.1 Division 1

4-3.1.1 Dust-Ignitionproof Equipment

The most common type of protection system used in Class II, Division 1 hazardous locations is dust-ignitionproof equipment. The *National Electrical Code* defines such equipment as "enclosed in a manner that will exclude ignitible amounts of dusts or amounts that might affect performance or rating and that, where installed and protected in accordance with [the *NEC*], will not permit arcs, sparks, or heat otherwise generated or liberated inside of the enclosure to cause ignition of exterior accumulations or atmospheric suspensions of a specified dust on or in the vicinity of the enclosure."

Unlike gases and vapors, dust can be kept out of equipment by conventional techniques, such as tight-fitting joints between mating parts of enclosures. The term "dust-ignitionproof" therefore includes a requirement that the equipment be dusttight, at least to the extent that the enclosure will exclude ignitible amounts of dust. However, it goes much beyond this.

The term requires that the equipment not permit arcs or sparks generated or liberated inside the enclosure to cause ignition of exterior accumulations or atmospheric suspensions of dust on or in the vicinity of the enclosure. The enclosure must, therefore, be constructed so that even if there is a fault resulting in spattered hot metal, the hot material and hot gases generated within the enclosure will not cause ignition of either a layer of dust on the outside of the enclosure or a cloud of dust around the enclosure. Since it has not proven practical to test for such conditions, the general rule of thumb used in standards covering dust-ignitionproof equipment is that the clearances between joints not exceed twice the minimum clearance specified for Group D explosionproof equipment and that the "flame path" (width of joint) not be less than 50 percent of the minimum width required for Group D explosionproof equipment. This is a general construction requirement established in recognized standards for dust-ignitionproof equipment, and it is not related to dust penetration tests. Such tests may demonstrate the need for tighter joints.

The definition also indicates that heat generated or liberated inside the enclosure should not cause ignition of exterior dust. This is basically the same requirement applied to explosionproof equipment. If the enclosure contains parts that operate at high temperatures, the external surface temperature of the enclosure should not exceed the ignition temperature of the dust. In addition, the thickness of the enclosure should be such as to prevent burn-through or hot spots as a result of arcing faults to the enclosure, at least for sufficient time to permit the circuit overcurrent devices to open the circuit. Thus, minimum enclosure thickness requirements are established.

The definition also indicates that the enclosure should be such that performance or rating of the equipment should not be affected by exterior dust. Heavy accumulations of dust are expected in Class II, Division 1 hazardous locations, and the dust acts as a thermal insulator. Therefore, the equipment should perform under normal conditions, even when radiation of heat from the enclosure is retarded by the thermal blanketing effect of the dust.

In addition to consideration of the layer and cloud ignition temperatures of the dust, the equipment should not operate at such a temperature that organic dust deposits will be changed through excessive dehydration or gradual carbonatization of deposits. Such changes can reduce the ignition temperature of the dust layer.

If the dust-ignitionproof equipment is rotating machinery, such as a motor, consideration also must be given to prevention of dust in bearings. Although the ignition from an overheated bearing is not electrical in nature, an improperly protected bearing on rotating electrical equipment can result in ignition just as readily as overheating of the equipment due to an electrical problem. This becomes even more critical if the dust is of an abrasive type, such as metal dust. If such a dust is inside a motor bearing, wearing and overheating of the bearing is almost inevitable.

A number of standards cover the construction and performance of dust-ignitionproof equipment, and many of them are recognized as American National Standards. Most of these standards are published by Underwriters Laboratories Inc.

In appearance, a dust-ignitionproof enclosure is similar to an explosionproof enclosure. In fact, many explosionproof enclosures have also been investigated as dust-ignitionproof enclosures. A dust-ignitionproof enclosure does not, however, have to be strong enough to withstand the internal pressures resulting from an explosion, since the protection method, by definition, prevents the combustible material from entering the enclosure. (See Figure 4-53.)

Explosionproof equipment, on the other hand, is not necessarily dust-ignitionproof. The equipment inside the enclosure may not operate properly if there is a dust layer on the equipment retarding heat transfer. Also, an explosionproof enclosure is not necessarily dusttight, particularly if there are bearings involved. The *National Electrical Code* specifically indicates that equipment and wiring defined as explosionproof is not to be required and is not acceptable in Class II locations unless approved for such locations.

Figure 4-53. *Dust-ignitionproof enclosure that is not explosionproof.*

As in explosionproof equipment, gaskets are not usually employed in dust-ignitionproof equipment, although they can be used if properly designed.

Wiring Methods: Since the wiring methods in Class II, Division 1 hazardous locations need not be designed to withstand the pressures of explosions, the requirements are not as stringent as they are in Class I, Division 1 hazardous locations. In addition to rigid metal conduit and intermediate metal conduit with threaded fittings, Type MI cable with approved fittings suitable for Class II locations, and Type MC cable listed for Class II locations (in industrial establishments only), and in addition to flexible connection fittings approved for Class II locations, liquidtight flexible metal conduit and flexible cord are permitted where flexible connections are needed. Fittings and boxes in which taps, joints, or terminal connections are made, or that are used in Group E locations (where dusts of a combustible electrically conductive nature are present), are required to be approved for Class II locations, usually interpreted as dust-ignitionproof or pressurized. However, if the box or fitting does not include taps, joints, or terminal connections, the fittings and boxes are not required to be dust-ignitionproof or pressurized in Group F and G locations. They are, however, required to be provided with threaded bosses for connection to threaded conduit or cable terminations and to have close-fitting covers with no openings (such as holes for attachments or screws) through which dust might enter or through which sparks or burning material might escape. For example, an ordinary dusttight box can be used in Class II, Division 1, Group F and G locations if it does not contain taps, joints, or terminal connections and if it has threaded bosses for threaded conduit or termination fitting connections.

Seals: Seals are not required in Class II locations for the same reasons they are required in Class I locations, because the protection methods either prevent dust from entering the enclosures or reduce the available ignition energy below that needed to cause ignition of combustible dusts. Therefore, there is no need for a seal to prevent propagation of flames from one part of the wiring system to another or to prevent transmittal of combustible dust from a Division 1 location to a less-hazardous or nonhazardous location.

There is, however, need for a seal or other installation method that will prevent dust from entering an otherwise dust-ignitionproof enclosure. The *NEC* therefore requires that, where a raceway provides communication between an enclosure that is required to be dust-ignitionproof and one that is not, there exist one of the following:

1. A permanent and effective seal in the raceway
2. A horizontal raceway not less than 10 ft (3.05 m) long

3. A vertical raceway not less than 5 ft (1.52 m) long and extending downward from the dust-ignitionproof enclosure

The 10-ft (3.05-m) horizontal raceway and 5-ft (1.52-m) vertical raceway below the dust-ignitionproof enclosure are judged as being long enough to permit any dust that enters the raceways to settle out of the air before it reaches the dust-ignitionproof enclosure. If the raceway provides communication between a dust-ignitionproof enclosure and an enclosure in a nonhazardous location, no seals or minimum lengths of raceway are required. (See Figures 4-54, 4-55, and 4-56.)

Testing Requirements: In addition to the usual tests to determine whether dust-ignitionproof equipment is suitable for use in nonhazardous locations, two basic types of tests are conducted.

Dust-penetration tests are conducted to determine that the equipment is dusttight. These tests are different from those conducted on dusttight equipment. The equipment is placed in a chamber in which a cloud of dust is circulated. The type of dust used depends upon the particular testing laboratory conducting the test and the standard used for evaluation of the equipment. Whichever type is used, the dust is very finely ground to a powder consistency. The requirements specify the sieve size through which the dust must pass if it is to be used in the test.

The equipment being tested, if it is heat-producing, is cycled on

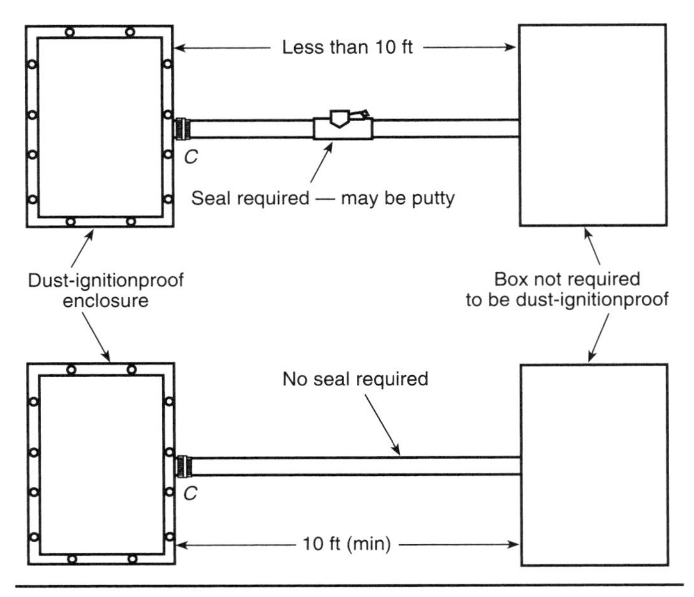

***Figure 4-54.** Illustration of requirements for seals in Class II locations in horizontal conduit runs. (Courtesy of HEP–Killark.)*

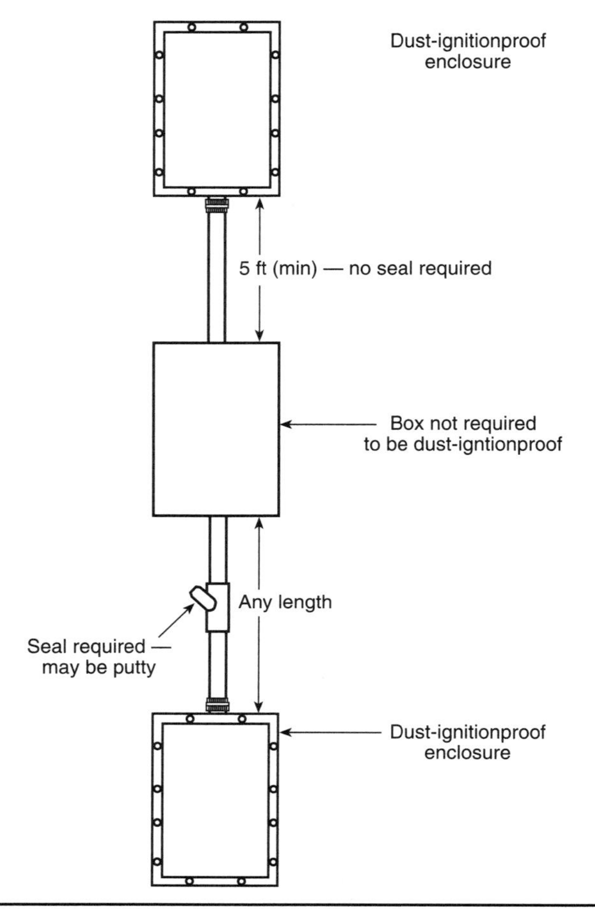

Figure 4-55. *Illustration of requirements for seals in Class II locations in vertical conduit runs. (Courtesy of HEP–Killark.)*

and off over a period of time in the atmosphere containing the circulating dust. Motors are tested under their rated load so that normal heating and cooling cycles are represented. Following this, the dust is carefully removed from the test chamber and the enclosure under test is opened to see if any dust has entered. In actual testing sequence, the dust-blanketing temperature test is usually conducted immediately following the dust penetration test and before the enclosure is opened.

The dust-blanketing temperature test is performed to determine that the external surface temperature of the equipment does not exceed the marked operating temperature (or to determine what the marked operating temperature or temperature range should be). The test also deter-

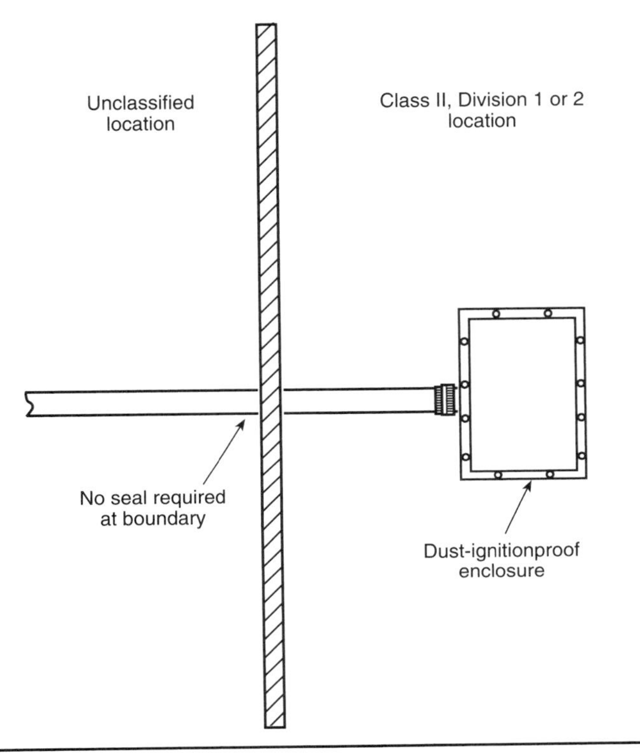

Figure 4-56. *Illustration of requirement for seals in Class II locations where conduit passes into unclassified location. (Courtesy of HEP–Killark.)*

mines that the equipment inside of the enclosure operates within the prescribed temperature limits, with the thermal blanket of the dust reducing heat transfer.

Installation and Maintenance Precautions: Installation and maintenance precautions for dust-ignitionproof equipment are essentially the same as for explosionproof equipment. The joints and gaskets should be protected during installation, bolts and nuts should be properly torqued, and threaded conduit systems should be made wrenchtight.

After the installation, the equipment should be checked for corrosion and poor connections. Ground continuity and adequacy also should be checked. If the enclosure is to be opened for any reason, such as maintenance, any dust that enters the enclosure during this operation should be carefully cleaned out (vacuuming with suitable equipment is recommended) before reclosing the enclosure.

Grounding and bonding is just as important in Class II, Division 1 locations as it is in Class I, Division 1 locations. Faults within the equipment must be cleared quickly to maintain the dust-ignitionproof properties of the enclosure and to prevent excessive surface temperatures that could ignite a dust layer.

4-3.1.2 Intrinsically Safe Equipment and Wiring

"Intrinsically safe" has exactly the same meaning in Class II locations as it does in Class I, Division 1 locations. The same standard is used for evaluating the equipment. (See Section 4-2.1.2.) The basic difference in the requirements for intrinsically safe equipment for Class II locations as compared to Class I, Division 1 locations is that spark ignition testing is not required if the circuit meets the requirements for intrinsically safe circuits for Class I locations. If the equipment is intended for use in Class II locations, testing or evaluation for either Group D locations or with methane is considered to represent testing with dusts. The reason for this rule is the difficulty in testing for spark ignition in dust atmospheres.

Because the ignition temperatures of dusts are generally below the ignition temperatures of gases and vapors, intrinsically safe equipment for Class II locations is limited to surface temperatures on exposed parts of 200°C (392°F) for Groups E and F locations, and 165°C (329°F) for Group G locations. An exception is made where tests show that higher temperature excursions on small parts will not result in ignition or charring of the appropriate test dusts.

If intrinsically safe apparatus for use in Class II locations is not enclosed in a dusttight enclosure, the standard requires that all bare energized parts within the enclosure should be assumed to be subject to fault between parts, with the number of faults unlimited and in the most unfavorable condition.

4-3.1.3 Pressurized Equipment

The *National Electrical Code* specifically indicates that pipe-ventilated motors are acceptable in Class II, Division 1 locations. This is a form of pressurized power equipment that is specifically recognized in NFPA 496, which includes the requirements for pressurized enclosures in Class II locations.

The Class II protection method equivalent to purged enclosures in Class I locations is somewhat different, although the principles are the same. (See Section 4-2.1.3.) If dust has entered the enclosure, an air "purging" operation is unlikely to remove the dust, at least that portion that has settled on the equipment. Therefore, the interior of the enclosure needs to be cleaned of dust before repressurization and reenergization of the equipment inside the enclosure is permitted. The word "purged" has no significance in Class II locations. Cleaning and then pressurizing without continuous flow is recognized for small enclosures not exceeding 10 ft^3 (0.3 m^3) in volume. For power equipment enclosures greater than

10 ft (0.3 m^3) in volume, either pressurization without continuous flow or, where the equipment requires air flow for heat dissipation, pressurization with continuous flow is permitted. In control rooms where people will be working, continuous-flow pressurization is permitted.

Because the extremely low pressures permitted for Class I purged equipment may be insufficient where high-density dust is present, the minimum pressure for pressurizing enclosures in Class II locations is 0.1 in. of water [0.0036 psi (25 Pa)], but only if the specific particle density is equal to or less than 130 lb/ft^3 (2083 kg/m^3). If the specific particle density exceeds these values, the minimum pressure permitted is 0.5 in. of water [0.018 psi (125 Pa)].

The temperature of the external surface of a pressurized enclosure in Class II locations is not permitted to exceed 80 percent of the layer-ignition temperature (in degrees Celsius) of the combustible dust involved, and in all cases is required to be at least 25°C (77°F) below this layer-ignition temperature. If the layer-ignition temperature of the dust involved is not known, the maximum surface temperature is limited in accordance with Table 4-3.

There are no levels of protection (such as Type X, Type Y, and Type Z) in the Class II pressurized-enclosure requirements. Other than the above, the requirements for pressurized equipment in Class II locations are essentially the same as they are for purged equipment in Class I locations as described in Section 4-2.1.3.

4-3.2 Division 2

4-3.2.1 Equipment for Division 1 Locations

In Class II, Division 2 locations, equipment suitable for Class II, Division 1 locations is acceptable. (See Section 4-3.1.)

Table 4-3 Maximum Surface Temperature

	Equipment Not Subject to Overloading		Equipment That May Be Overloaded*			
			Normal Operation		Abnormal Operation	
Class II Group	°C	°F	°C	°F	°C	°F
E	200	392	200	392	200	392
F	200	392	150	302	200	392
G	165	329	120	248	165	329

For example, motors or power transformers.

4-3.2.2 Dusttight Equipment

For most equipment in Class II, Division 2 locations, the *National Electrical Code* indicates that dusttight equipment can be used.

Dusttight is defined in the *National Electrical Code* as "so constructed that dust will not enter the enclosing case under specified test conditions." ANSI/NEMA 250[27] indicates that either of two different test methods can be used: a dust-blast test method or an atomized water test method.

NEMA and UL designations for dusttight equipment are Types 3, 3S, 4, 4X, 6, 6P, 12, 12K, and 13. NEMA Type 9 enclosures (dust-ignitionproof) are also acceptable.

4-3.2.3 Nonincendive

Nonincendive equipment has the same definition and essentially the same requirements for Class II, Division 2 locations as it does for Class I, Division 2 locations. (See Section 4-2.2.2.) However, since it is relatively easy to find and install dusttight equipment and enclosures, the nonincendive protection method is not common in Class II, Division 2 locations.

4-3.2.4 Pressurized Enclosures

The same types of pressurized enclosures recognized for Class II, Division 1 locations can be used in Class II, Division 2 locations. (See Section 4-3.1.3.) Pressurization of electrical equipment rooms in Class II, Division 1 and Division 2 locations (for example, grain elevators) is quite common.

4-3.2.5 Other Types

Other types of protection systems are also recognized in Class II, Division 2 locations. Pipe-ventilated motors similar to those recognized in Class II, Division 1 locations are recognized. For some specific types of equipment, tight-fitting or telescoping joints with no openings in the enclosure but not necessarily "dusttight" by definition are also recognized.

4-3.2.6 Wiring Methods

In Class II, Division 2 locations, all of the wiring methods recognized in Class II, Division 1 locations are acceptable. Although cable trays are not recognized in Class II, Division 1 locations (because they are considered "dust collectors" and difficult to keep clean), they are recognized in Class II, Division 2 locations if the cable is installed in a ventilated channel-type cable tray in a single layer with a space of not less than the larger cable diameter between any two adjacent cables. Other recognized wiring methods include several different types of cables, dusttight wireways, and electrical metallic tubing (EMT).

Wireways, fittings, and boxes in which taps, joints, or terminal connections are made are not required to be dusttight if they have no openings through which, after installation, sparks or burning material might escape or through which adjacent combustible material might be ignited. They must also be provided with telescoping or close-fitting covers or other effective means to prevent the escape of sparks or burning material.

4-4 Class III Locations

4-4.1 Divisions 1 and 2

4-4.1.1 Equipment Suitable for Class II, Division 2

Equipment suitable for Class II, Division 2 locations is acceptable for use in Class III locations.

4-4.1.2 Special Requirements

The *National Electrical Code* includes special requirements for some equipment, such as lighting fixtures, motors, cranes and hoists, and plugs and receptacles, applicable only to equipment installed in Class III locations.

Electric lighting fixtures not suitable for Class II, Division 2 locations may be of a type designed to minimize the entrance of fibers and flyings and to prevent the escape of sparks, burning material, or hot metal.

Where, in the judgment of the authority having jurisdiction, only moderate accumulations of lint or flyings will be likely to collect on or in the vicinity of rotating electric machines, and where such a machine is readily accessible for routine cleaning and maintenance, the following types of machines are permitted:

1. Self-cleaning textile motors of the squirrel-cage type
2. Standard open-type machines without sliding contacts, centrifugal, or other types of switching mechanisms, including motor overload devices
3. Standard open-type machines having switching contacts and the like, provided they are enclosed in a tight housing without ventilation or other openings.

Electric cranes, hoists, and similar equipment, where installed for operation over Class III locations (including traveling cranes and hoists for material handling, traveling cleaners for textile machinery, and similar equipment), are required to have the power supply to the contact conductors isolated from all other systems. The power supply must also be equipped with a ground detector that either will give an alarm and automatically deenergize the contact conductors in case of a fault to ground or will give a visual and audible alarm as long as power is supplied to the contact conductors and the ground fault remains. The contact conductors are required to be located or guarded so as to be inaccessible to other-than-authorized persons, and they are required to be protected against accidental contact with foreign objects. Current collectors are required to be so arranged or guarded as to confine normal arcing and sparking and to prevent the escape of sparks or hot particles. At least two separate surfaces of contact are required to be provided for each contact conductor in order to reduce sparking. The *NEC* also requires that reliable means be provided to keep contact conductors and current collectors free of accumulations of lint or flyings.

Receptacles and attachment plugs in Class III locations are required to be of the grounding type and designed so as to minimize the accumulation or the entry of fibers or flyings and to prevent the escape of sparks or molten particles. However, in locations where, in the judgment of the authority having jurisdiction, only moderate accumulations of lint or flyings are likely to collect in the vicinity of a receptacle, and where the receptacle is readily accessible for routine cleaning, general-purpose grounding-type receptacles mounted so as to minimize the entry of fibers or flyings are permitted.

There are also special rules for ventilating piping in Class III locations. These include requirements for seams, minimum metal thickness, location and protection of the inlet opening, and protection against damage.

4-4.1.3 Surface Temperatures

External surface temperatures under operating conditions are limited to 165°C (329°F) for equipment, such as electric lighting fixtures, that is not subject to overloading, and 120°C (248°F) for equipment that may be overloaded, such as motors and power transformers.

4-4.1.4 Wiring Methods

In Class III, Division 1 locations, wiring methods are required to be rigid metal conduit, rigid nonmetallic conduit, intermediate metal conduit, electrical metallic tubing, dusttight wireways, Type MI cable, or Type MC cable. Boxes and fittings are required to be dusttight where flexible connections are necessary; the requirements are the same as for Class II, Division 2 locations.

In Class III, Division 2 locations, the same requirements apply, with the following exception: in sections, compartments, or areas used solely for storage or containing no machinery, open wiring on insulators is permitted where installed in accordance with the requirements for such wiring — but only under certain conditions where the conductors are protected from physical damage.

Bibliography

[1]NFPA 70, *National Electrical Code®*, National Fire Protection Association, Quincy, MA, 1996.

[2]NEMA 250, *Enclosures for Electrical Equipment (1000 Volts maximum)*, National Electrical Manufacturers Association, 1300 N. 17th Street, Suite 1847, Rosslyn, VA, 1991.

[3]ANSI B46.1, *Surface Texture (Surface Roughness, Waviness, and Lay)*, American National Standards Institute, New York, NY, 1985.

[4]*Hazardous Locations Equipment Directory* (PTDR), Underwriteres Laboratories Inc., 333 Pfingsten Road, Northbrook, IL, 1997.

[5]ANSI/UL 913, *Standard for Intrinsically Safe Apparatus and Associated Apparatus for Use in Class I, II, and III, Division 1 Hazardous (Classified) Locations*, Underwriters Laboratories Inc., Northbrook, IL, 1988.

[6]ANSI/ISA RP12.6, *Installation of Intrinsically Safe Instrument Systems in Class I Hazardous Locations*, Instrument Society of America, Research Triangle Park, NC, 1987.

[7]NFPA 496, *Standard for Purged and Pressurized Enclosures for Electrical Equipment,* National Fire Protection Association, Quincy, MA, 1993.

[8]ANSI/ISA S12.12, *Electrical Instruments for Use in Class I, Division 2 Hazardous (Classified) Locations*, Instrument Society of America, Research Triangle Park, NC, 1984.

[9]UL 1604, *Electrical Equipment for Use in Hazardous Locations, Class I and Class II, Division 2 and Class III, Divisions 1 and 2*, Underwriters Laboratories Inc., Northbrook, IL, 1994.

[10]ANSI/UL 844, *Electric Lighting Fixtures for Use in Hazardous (Classified) Locations*, Underwriters Laboratories Inc., Northbrook, IL, 1995.

[11]ANSI/UL 698, *Industrial Control Equipment for Use in Hazardous (Classified) Locations,* Underwriters Laboratories Inc, Northbrook, IL, 1991.

[12]IEC 79-11, *Electrical Apparatus for Explosive Gas Atmospheres Part*

11: Intrinsic Safety "i," International Electrotechnical Commission, Geneva, Switzerland, 1991.

[13]IEC 79-3, *Electrical Apparatus for Explosive Gas Atmospheres Part 3: Spark Test Apparatus for Intrinsically-safe Circuits,* International Electrotechnical Commision, Geneva, Switzerland, 1990.

[14]IEC 79-1, *Electrical Apparatus for Explosive Gas Atmospheres Part 1: Construction and Verification Test of Electrical Apparatus* and *Amendment No. 1 (1993),* International Electrotechnical Commission, Geneva, Switzerland, 1990.

[15]ISA S12.22.01, *Electrical Apparatus for Use in Class I, Zone 1 Hazardous (Classified) Locations, Type of Protection — Flameproof "d,"* Instrument Society of America, Research Triangle Park, NC, 1996.

[16]UL 2279, *Electrical Equipment for Use in Class I, Zone 0, 1, and 2 Hazardous (Classified) Locations*, Underwriters Laboratories Inc, Northbrook, IL, 1996.

[17]IEC 79-7, *Electrical Apparatus for Explosive Gas Atmospheres Part 7: Increased Safety "e,"* and *Amendment No. 1 (1991),* and *Amendment No. 2 (1993),* International Electrotechnical Commission, Geneva, Switzerland, 1990.

[18]ISA S12.16.01, *Electrical Apparatus for Use in Class I, Zone 1 Hazardous (Classified) Locations, Type of Protection — Increased Safety "e,"* Instrument Society of America, Research Triangle Park, NC, 1996.

[19]IEC 79-2, *Electrical Apparatus for Explosive Gas Atmospheres Part 2: Electrical Apparatus, Type of Protection "p,"* International Electrotechnical Commission, Geneva, Switzerland, 1983.

[20]IEC 79-13, *Electrical Apparatus for Explosive Gas Atmospheres Part 13: Construction and Use of Rooms or Buildings Protected by Pressurization,* International Electrotechnical Commission, Geneva, Switzerland, 1982.

[21]IEC 79-6, *Electrical Apparatus for Explosive Gas Atmospheres, Part 6: Oil Immersion "o,"* International Electrotechnical Commission, Geneva, Switzerland, 1995.

[22]ISA S12.26.01, *Electrical Apparatus for Use in Class I, Zone 1 Hazardous (Classified) Locations, Type of Protection Oil Immersion "o,"* Instrument Society of America, Research Triangle Park, NC, 1996.

[23]IEC 79-5, *Electrical Apparatus for Explosive Gas Atmospheres Part 5: Powder Filling, Type of Protection "q,"* International Electrotechnical Commission, Geneva, Switzerland, 1996.

[24]ISA S12.25.01, *Electrical Apparatus for Use in Class I, Zone 1 Hazardous (Classified) Locations—Powder Filling "q,"* Instrument Society of America, Research Triangle Park, NC, 1996.

[25]IEC 79-18, *Electrical Apparatus for Explosive Gas Atmospheres Part 18: Encapsulation "m,"* International Electrotechnical Commission, Geneva, Switzerland, 1992.

[26]ISA S12.23.01, *Electrical Apparatus for Use in Class I, Zone 1 Hazardous (Classified) Locations Type of Protection — Encapsulation "m,"* Instrument Society of America, Research Triangle Park, NC, 1996.

[27]ANSI/NEMA 250, *Enclosures for Electrical Equipment (1000 Volts Maximum),* National Electrical Manufacturers Association, Washington, DC, 1985.

CHAPTER 5

Static Electricity and Lightning Protection

5-1 Static Electricity

Static electricity and lightning are two sources of ignition for flammable and combustible liquids, dusts and fibers, and flammable gases.

The term "static electricity" is used to mean the electrification of materials through physical contact and separation, and the effects of the positive and negative charges so formed, particularly where sparks may result that constitute a fire or explosion hazard.

In order for static electricity to be a potential ignition source, the following four conditions must be met:

1. There must be an effective means to generate a static electric charge.
2. There must be a means of accumulating the separate charges and maintaining a suitable difference of electrical potential.
3. There must be a discharge of the accumulate charge, of adequate energy.
4. The spark must occur in an ignitible mixture.

Static electricity is generated when two bodies are in close contact with each other. Free electrons are transferred between these two bodies, and an attractive force is established between them. When these bodies are separated, work must be done in opposition to these attractive forces. This results in the objects becoming charged with negative and positive charges, with one body becoming negatively charged and the other body becoming equally charged positive. This creates a potential difference between the charged bodies and also between the charged bodies and ground. If there is a conductive path between these two bodies, the charges will reunite and the objects will become electrically neutral. If there is no conductive path between the bodies, the charge remains on the bodies. Eventually, this charge will dissipate as it slowly "bleeds" off by ionization of the surrounding air. The charge can also dissipate to ground if a conductive path exists between the charged bodies and ground.

Induction also plays a role in static electricity. Like charges repel each other, while opposite charges attract. If a charged insulator is brought in close proximity to a conductor, the charged surface of the insulator will attract an equal and opposite charge on the conductor. A charge of opposite polarity will be repelled to the opposite end of the conductor. (This charge would have the same polarity as the charge on the insulator.) This is called induction. The charge on the end of the conductor that is nearest to the insulator is bound, while the charge on the opposite end is free to dissipate. If this end of the charged conductor is brought momentarily to ground, the free charge can dissipate, and the conductor will have a net excess charge. When the insulator is moved away from the conductor, the bound charge is then free to redistribute over the entire surface of the conductor. This charge now has the potential of releasing a spark. In many cases, the induced charges are far more dangerous than a charge on an insulator. Induced charges can be released in their entirety in a single spark, while an insulator can only release a portion of its charge from a small area.

The energy of the spark can be found by knowing the capacitance and voltage of the objects and the spark, respectively. The relationship is shown as follows:

$$\text{Energy (in joules)} = C(V)^2$$

where: C = capacitance (farads)
V = potential difference (volts)

Typical values of energy would be in millijoules.

In many cases, one of the two plates of the capacitor is the earth, while the insulator between the two plates is air.

Tests have shown that optimum mixtures of hydrocarbons, gases, and vapors in air require approximately 0.25 mJ of energy for ignition to occur. Unsaturated hydrocarbons may have lower ignition energies. Some typical values are shown in Table 5-1. See also Section 2-8.1.

The ignition energy of optimum mixtures of dust and fibers with air is about one or two magnitudes greater than those of gases and vapors. (See Section 2-8.1.)

Static electricity is a potential hazard in areas where flammable liquids, flammable gases, and combustible dusts and fibers are present. In flammable liquids, the electrical conductivity of the liquid plays a major role in the ability of the liquid to hold a charge. If a liquid has a conductivity greater than 50 picoseimens/meter (resistivity less than 200 megaohm-centimeter), any charge generated will dissipate before reaching a hazardous potential.

Table 5-1 Approximate Minimum Ignition Energy

Material	Minimum Ignition Energy (millijoules)
Methane	0.30
Propane	0.25
Cyclopropane	0.18
Ethylene	0.08
Acetylene	0.017
Hydrogen	0.017

Liquids stored in tanks (in most cases metal tanks) are either in contact with earth, concrete, or other slightly conducting foundation, or they are insulated from earth, such as in a tank vehicle with dry rubber tires. In both cases, if an electrically charged liquid is poured into a tank or container, the like charges are repelled from each other inside the liquid. These charges are repelled toward the outer edges of the liquid. The charge on the surface is called the "surface charge," which is of concern in most cases.

In the first scenario, when the metal container or tank is connected with the earth, the tank is considered electrically neutral. The charged liquid in the tank will have a surface charge. This surface charge will attract an opposite charge on the tank, which is connected to earth. Eventually, the opposite charges will reunite and the charged liquid becomes neutral. This bleeding-off or reunion of the charge is called relaxation time and is inversely related to the conductivity of the liquid. If a liquid is not very conductive, the relaxation time is longer. The relaxation time could be from a fraction of a second to several minutes, depending on the electrical conductivity of the liquid.

If the potential difference between the liquid surface and metal tank is high enough, ionization of the air can occur and a spark can jump to the tank. If a flammable vapor-air mixture is present, a fire or explosion will occur.

In the second scenario, the tank is insulated from ground, for example, a tank truck insulated by dry rubber tires. When the tank is being filled with a liquid, a charge develops on the surface. This surface charge will attract a charge of opposite polarity on the interior of the tank wall. The exterior of the tank will have a free charge of the same polarity as the surface charge of the liquid. This charge is then capable of producing a spark to ground. On a tank truck, this spark could be between the open dome and a grounded fill pipe.

Static electricity can be controlled in industrial processes and procedures. Several methods can be used to dissipate a static charge or prevent its accumulation.

There are many types of materials that are ordinarily thought of as insulators, such as paper, fabric, and wood. These same materials can have a conductive surface if subjected to an atmosphere that is humid. These insulating materials usually will not build up an appreciable static electric charge when they are in an atmosphere with a high relative humidity. The surface of the insulators will be slightly conductive because of the moisture layer, and this layer allows the charge to bleed off and not accumulate.

Many times static electricity seems to be more of a problem in the winter. This is because the outside air, which may have a high relative humidity, is brought indoors and heated. When the air is heated, the relative humidity decreases. As an example, saturated air at an outdoor temperature of 30°F (-1°C) would have a relative humidity of only a little over 20 percent when brought indoors and heated to 70°F (21°C).

In general, a relative humidity of 50 percent or greater will allow charges to bleed off and not accumulate. A humidifier will help in most cases. However, in some cases, such as an oily surface that will not absorb the moisture, it will not help. In situations where natural conditions such as relative humidity do not allow the static charge to bleed off, this must be done artificially, by bonding or grounding.

Grounding (also known as "earthing") is the process of connecting a conductive object to ground or earth by means of a conductive path, such as a wire or cable. Grounding helps ensure that any charge accumulating on the object can easily bleed to ground, thus removing the excess on the object. Grounding ensures that the objects are at the same potential as the earth. Bonding is the process of connecting two or more conductive objects together by a conductive path, so they will all be at equal, but not necessarily zero, potential. If one object is grounded, all others connected to it are also grounded. Some objects are inherently grounded, such as water pipes or a tank resting on the earth.

Bonding or grounding can be done with wires, cables, or other conductive media. The size of the wire must be capable of carrying the current. Since static electric currents are small, in the order of microamperes, the wire size is not relevant. However, the wire must be heavy enough to provide mechanical strength, especially if repeated connection, disconnection, and movement are the norm. In addition, the resistance can be 1 megaohm or higher to bleed off a static charge. It may be lower, depending on the hazard.

Bonding and grounding are used on tank trucks and containers when they are being filled with a flammable liquid. As an example, the tank truck would be bonded to the loading pipe or directly to a ground. If the tank truck is grounded, the loading pipe should also be grounded. This will equalize the potential between the tank truck and loading pipe. (See Figure 5-1.)

Safety precautions for filling metal containers with flammable

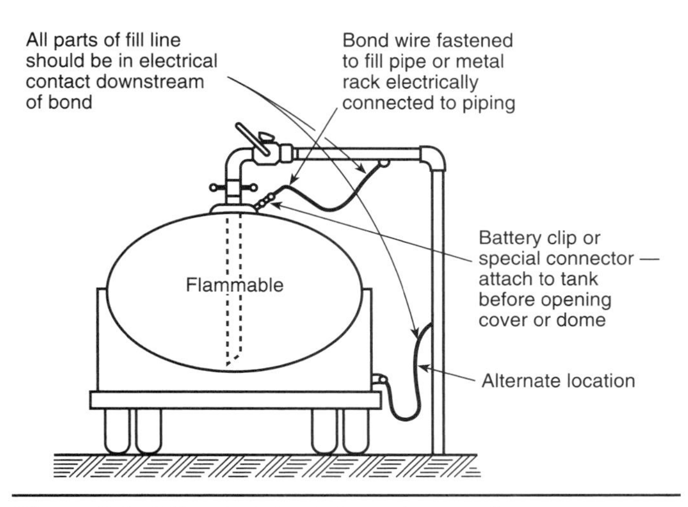

Figure 5-1. *Filling tank truck through open dome.*

liquids are less rigid. The filling nozzle should be kept in contact with the container. This will equalize the potential between the container and the pipe. Other methods of bonding are shown in Figure 5-2.

Another way for a charge to bleed off is through air that is made conductive by ionization. Ionization can be accomplished by the use of a static comb, electrical neutralization, or radioactive neutralizer.

A static comb is a metal bar equipped with a series of needle points or a metal wire surrounded by metallic tinsel. On a conducting body, the static charge is free to flow. On a spherical body, the charge will distribute evenly.

The charge on a nonspherical body will concentrate on the surface having the least radius of curvature, such as a point. If the point is surrounded by air, the charge concentration at the point will ionize the air, and a charge will be able to flow through the conductive air. When the grounded static comb is brought next to a charged object, the air is ionized at the points, and a charge flows between the static comb and the charged object. Static combs are used in neutralizing charged fabrics, power belts, and paper.

Electrical neutralizers utilize a line-powered, high-voltage source to ionize the air. This device is used for removing static charges from cotton, wool, silk, and paper in process and printing.

Radioactive neutralizers use a radioactive source to ionize the air. Some problems associated with this device include health hazards, dust accumulation, ensuring proper positioning, and determining the amount of radiation required.

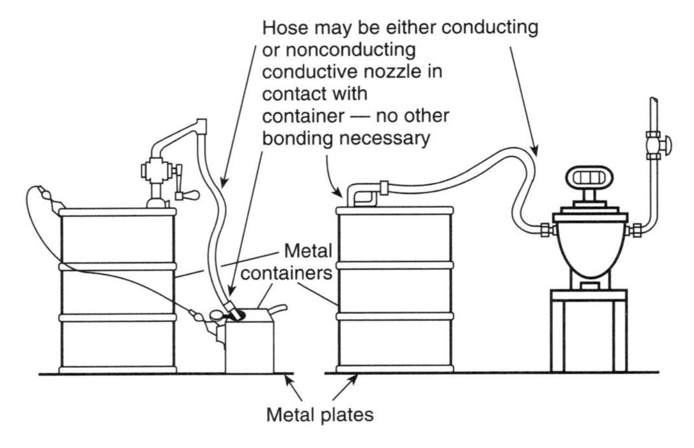

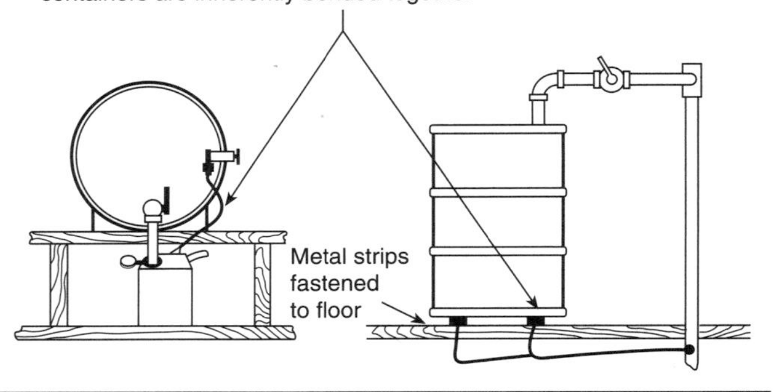

Figure 5-2. *Bonding during container filling.*

For further information on static electricity, refer to NFPA 77, *Recommended Practice on Static Electricity*[1].

5-2 Lightning Protection

Lightning is frequently the primary cause of fires and is often an indirect cause of fires and explosions because of induced voltages and sparks, which may ignite flammable vapors.

There are four different types of lightning strokes as follows:

1. The negative downward stroke
2. The positive downward stroke
3. The positive upward stroke
4. The negative upward stroke

Lightning does not always occur between clouds and the ground. Lightning strokes can also occur between clouds.

Lightning is generated in clouds. These clouds are very large and can reach a height of over 65,000 ft (19,825 m). The bases are usually from 0.5 mi (0.8 km) to 2.5 mi (4 km) high and from 5 mi (8.04 km) to 30 mi (48.27 km) in diameter. These clouds contain water droplets, ice particles, snow, and hail. Clouds such as these can produce as much as 1.5 in. (38 mm) of rain in a very short time over an area of at least 1 mi^2 (2.6 km^2). This much water would weigh over 100,000 tons (90.7 $\times 10^6$kg). It takes upward-moving winds of considerable speed to keep this much moisture suspended in air.

This high, upward air movement in a thundercloud also moves the hailstones, water droplets, ice particles, and snow inside the clouds. The movement of these particles against each other leads to a charge buildup inside the cloud, similar to the way static electricity is generated. The charges tend to separate, with the negative charges on the bottom of the cloud and the positive charges on the top of the cloud. As the cloud moves over the earth and objects on the earth, the negative charge on the bottom of the cloud attracts a positive charge on earth and its objects. Eventually, a downward leader is initiated from the cloud. It progresses toward the earth in discrete steps. After each step, the leader pauses for about 50 microseconds and then proceeds downward again on a different path. As the leader heads downward, it ionizes the air and thus allows a current to flow. The tip of the leader is essentially at the same potential as the cloud. At some distance from the earth, known as the striking distance, the gap is completed, and this is known as the point of discrimination. The return stroke from earth contains the major part of the lightning discharge. As many as 40 component strokes have been observed from a single flash. The speed ranges from 100 mi/second (161 km/second) for the leader stroke to 20,000 mi/second (32,000 km/second) for the main stroke. The current can be as high as 270,000 A in the most extreme case, lasting for a few microseconds or 10,000 A in a more typical case, lasting for a longer duration. The potential can be as high as 15 million V.

Tanks containing flammable and combustible liquids and gases stored at atmospheric pressure have been known to be set on fire by lightning. Fires can be started by a direct hit from lightning or from sparking induced from a lightning strike in the vicinity of the tank.

On ordinary structures, fires can start from a direct strike. They also can start as a result of damage to electrical equipment when high currents enter the building on the electrical service, either because overhead wires are struck or because very high currents are induced in these overhead conductors by a nearby strike.

At the present, NFPA 780, *Standard for the Installation of Lightning Protection Systems*[2], does not address the protection of buildings containing explosive materials or munitions. Further information on this subject can be obtained from the following locations: Naval Publications and Forms Center, Philadelphia, PA; Headquarters, Army Material Command Code DRXAM-ABS, Alexandria, VA; or Air Force Publications Center, Baltimore, MD, as well as in the following publications:

DoD 6055.9-STD, *Ammunition and Explosives Safety Standard*[3]

NAVSEA OP-5, *Ammunitions and Explosives Ashore*[4]

AMCR 385-100, *Safety Manual*[5]

AFR 127-100, *Explosives Safety Standards*[6]

MIL-HDBK-419, *Grounding, Bonding, and Shielding for Electronic Equipment and Facilities*[7]

The following tanks or vessels containing flammable liquids and gases are in effect protected against lightning and need no additional protection:

1. Metal structures that are electrically continuous
2. Tanks or vessels that are tightly sealed to prevent the escape of liquids, vapors, or gases
3. Tanks or vessels that are 0.1875 in. (4.77 mm) or more in thickness

If the tanks do not meet the above requirements, then the following should be adhered to for protection of these structures and their contents from lightning damage:

1. Flammable liquids must be stored in essentially gastight structures.
2. Openings where flammable concentrations of vapor or gas can escape to the atmosphere must be closed or otherwise protected against the entrance of flame.
3. Structures and all appurtenances (for example, gauge hatches and vent valves) must be maintained in good operating condition.
4. Flammable vapor-air mixtures must be prevented, to the greatest possible extent, from accumulating outside such structures.
5. Potential spark gaps between metal conductors must be avoided at points where flammable vapors may escape or accumulate.

In addition, the lightning-protection system could employ air terminals (lightning rods), masts, or overhead wires. The most common type of protection on ordinary structures consists of air terminals, down conductors, and ground rods or a loop conductor system. Loop conductor systems will be examined in more detail later in this chapter.

The mast and overhead-wire method gives a zone of protection. Both masts and overhead-wire protection are based on a striking distance of 100 ft (30.5 m). In most cases, the striking distance (the distance over which final breakdown of the initial strike to ground or to a grounded object occurs) exceeds 100 ft (30.5 m). Therefore, the zone of protection is based on 100 ft (30.5 m) to give an adequate level of protection. The zone of protection is a circular arc concave upward from the top of a mast or from an overhead wire. (See Figure 5-3.)

Fires on tanks with floating roofs are often caused by lightning around the vapor seal. This can be from a direct stroke or induced sparks. The roofs of these tanks should be bonded to the shoes of the seals at intervals not exceeding 10 ft (3.05 m). This will equalize the potential between the tank walls and the roof, thus reducing the possibility of a spark jumping the gap between the roof and tank where a flammable vapor would be.

Ordinary structures are protected with a system containing air terminals, down conductors, and grounding. In ordinary structures, processes and environments may exist that could be classified as hazardous. When lightning strikes a building, there is the potential for flashover or sideflash to occur, thus igniting a flammable liquid, vapor, or dust. Flashover or sideflash is when an arc jumps between two objects at different potential. For instance, when lightning strikes an air terminal, current flows down a conductor to ground. There is resistance and inductance on the down conductor. This leads to a potential on the down conductor. When a grounded metal object is in close proximity to this conductor, an arc or flashover can occur between the down conductor and the grounded metal object.

A standard lightning protection system for ordinary structures consists of an air terminal, a ground system, and conductors connecting the two. The theory is that this gives the lightning current a low resistance path to ground.

When the lightning current follows through a high-resistance path, damage will usually result because of the heat and mechanical forces. A high-resistance path is wood, brick, tile, stone, concrete, or other similar material.

The parts of a building most likely to be struck by lightning are those that project above the surrounding parts, such as chimneys, ventilators, flagpoles, towers, water tanks, spires, gables, skylights, dormers, ridges, and parapets. On a flat roof, the edge is most likely to get struck. The air terminals can be spaced every 20 ft (6.1 m) for an air terminal of not less than 10 in. (254 mm) and every 25 ft (7.62 m) for an air terminal of not less than 24 in. (610 mm) above the object being protected on a flat roof. A zone of protection is developed by the air terminals for a striking distance of 150 ft (45.7 m). On flat roofs that exceed a length or width of 50 ft (15.2 m), additional air terminals must be provided.

All air terminals must have two paths to ground, except for some situations that involve a dormer or other similar structure. Down conduc-

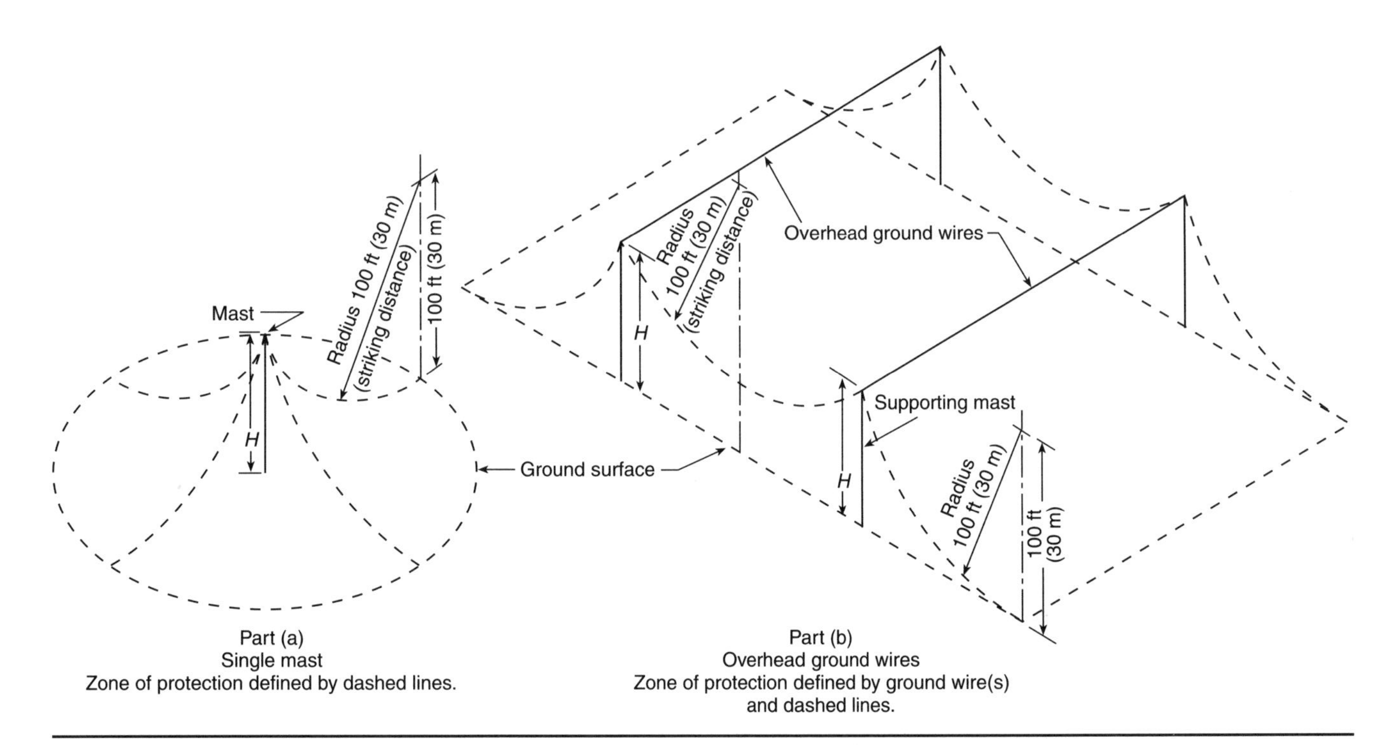

Figure 5-3. *Everything under the arc, known as the zone of protection, is protected.*

tors connect the air terminal to a ground system. There should always be at least two down conductors for every system. Structures exceeding 200 ft (61 m) in perimeter must have a down conductor for every 100 ft (30.5 m) of perimeter or fraction thereof.

The grounding of the system is very important. The objective is to obtain a low-resistance ground. The typical grounding system consists of ground rods. Ground rods are copper-clad steel, solid copper, or stainless steel not less than 0.5 in. (12.7 m) in diameter and 8 ft (2.44 m) in length. The ground rod must be not less than 10 ft (3.05 m) into the earth. There are other grounding methods that can be used to obtain better grounds for high-resistance soil conditions. These include multiple ground rods and buried grid systems.

Bonding of metal bodies and grounded metal bodies to the lightning protection system is needed to reduce the possibility of a sideflash. The bonding of these metal bodies will equalize the potential between the two metal objects. For further information on lightning protection, refer to NFPA 780.

Bibliography

[1]NFPA 77, *Recommended Practice on Static Electricity,* National Fire Protection Association, Quincy, MA, 1993.

[2]NFPA 780, *Standard for the Installation of Lightning Protection Systems,* National Fire Protection Association, Quincy, MA, 1997.

[3]DoD 6055.9-STD, *Ammunition and Explosives Safety Standards,* Chapter 7, Department of Defense, Washington, DC, July 1984.

[4]NAVSEA OP-5, *Ammunitions and Explosives Ashore,* Vol. 1, fourth revision, Chapter 4, Naval Sea Systems Command, Washington, DC, May 1983.

[5]AMCR 385-100, *Safety Manual,* Chapter 8, Army Material Command, Washington, DC, 1985.

[6]AFR 127-100, *Explosives Safety Standards,* Department of the Air Force, Washington, DC, May 1983.

[7]MIL-HDBK-419, *Grounding, Bonding, and Shielding for Electronic Equipment and Facilities,* Vols. I and II, Department of Defense, Washington, DC, January 1982.

CHAPTER 6

Requirements Outside the United States

Among the various international, regional, foreign national, and U.S. standards for hazardous locations, there are differences in construction, types of protection permitted, and installation requirements. However, in every case these standards are aimed at providing safe installations. This chapter will identify the various standards and provide an overview of their major differences.

6-1 Major Organizations

6-1.1 International Electrotechnical Commission

Many organizations outside the United States prepare standards for electrical equipment used in hazardous locations. The major organization for worldwide electrical standards is the International Electrotechnical Commission (IEC), which is based in Geneva. This organization was founded in 1906 and has grown so that there are now over 50 member countries. They are responsible for the electrical standardization activities. The International Standards Organization (ISO), founded in 1947, has the task of preparing the mechanical standards. With the rapid expansion of computer technologies, the IEC and ISO have established joint activities to expedite this work.

The IEC committee directly responsible for all hazardous location equipment is Technical Committee 31, Electrical Apparatus for Explosive Atmospheres. T/C 31 deals directly with several standards for types of protection. Also under this committee are eight subcommittees that deal with specific types of construction or classification of areas and installation rules.

These committees are the following:

1. S/C 31A — Flameproof enclosures
2. S/C 31C — Increased safety apparatus
3. S/C 31D — Pressurization and associated techniques
4. S/C 31G — Intrinsically safe apparatus
5. S/C 31H — Apparatus for use in the presence of ignitable dust
6. S/C 31J — Classification of hazardous areas and installation requirements
7. S/C 31K — Encapsulation
8. S/C 31L — Electrical apparatus for the detection of flammable gases

The IEC standards are printed and issued by IEC headquarters in Geneva. These standards have no official standing until they are adopted by a national standards organization.

The standards that have been developed by T/C 31 and its subcommittees are shown below and are available from

American National Standards Institute (ANSI)
11 West 42nd Street, 13th floor
New York, NY 10036 (USA)
telephone (212) 642-4900
fax (212) 302-1286
http://www.ansi.org.

The 60079 series of standards that follow are titled *Electrical Apparatus for Explosive Gas Atmospheres:*

60079-0 — General requirements

60079-1 — Construction and verification test of flameproof enclosures of electrical apparatus

60079-1A — Appendix D: Method of test for ascertainment of maximum experimental safe gap

60079-2 — Electrical apparatus, type of protection "p"

60079-3 — Spark-test apparatus for intrinsically-safe circuits

60079-4 — Method of test for ignition temperature

60079-5 — Powder filling "q"

60079-6 — Oil immersion "o"

60079-7 — Increased safety "e"

60079-10 — Classification of hazardous areas

60079-11 — Intrinsic safety "i"

60079-12 — Classification of mixtures of gases or vapors with air according to their maximum experimental safe gaps and minimum igniting currents

60079-13 — Construction and use of rooms or buildings protected by pressurization

60079-14 — Electrical installations in hazardous areas (other than mines)

60079-15 — Electrical apparatus with type of protection ‘‘n’’

60079-16 — Artificial ventilation for the protection of analyzer houses

60079-17 — Inspection and maintenance of electrical installations in hazardous areas (other than mines)

60079-18 — Encapsulation ‘‘m’’

60079-19 — Repair and overhaul for apparatus used in explosive atmospheres (other than mines or explosives)

60079-20 — Data for flammable gases and vapors, relating to the use of electrical apparatus

Another standard, under the auspices of T/C 70 instead of T/C 31, but nonetheless important in the application of the above listed standards, is IEC 60529, *Classification of Degrees of Protection Provided by Enclosures*[1]. This standard is discussed later in Section 6-2.3.

NOTE: In 1997, all IEC standards were renumbered by adding ‘‘60000’’ to the existing base numbers.

Also under the auspices of IEC is the Ex scheme, begun in 1995. It is an arrangement by which the country members, who must agree to accept the above IEC standards, may have the equipment certifications issued by their national testing houses (or ‘‘notified bodies’’) accepted by all other members.

6-1.2 CENELEC

A major regional standards organization is the result of the European Common Market. The Commission of the European Communities issued a directive in the mid-1970s to harmonize standards for electrical equipment for use in hazardous locations. The purpose of this directive was to facilitate trade within the community through the adoption of harmonized standards. The CENELEC Organization was assigned this task and was directed to base their new standards on the IEC work. This work was completed and standards were published. There are presently 18 national members of CENELEC. The resulting CENELEC standards and subsequent revisions are similar to the IEC standards, but there are differences. In several instances, CENELEC standards have been proposed as revised IEC standards, which resulted in very similar documents. Nevertheless, it is recommended that one obtain the CENELEC standards if they must be followed.

The CENELEC standards are listed as follows:

EN 50014 — Electrical apparatus for potentially explosive atmospheres — General requirements

EN 50015 — Oil immersion "o"

EN 50016 — Pressurized apparatus "p"

EN 50017 — Powder filling "q"

EN 50018 — Flameproof enclosure "d"

EN 50019 — Increased safety "e"

EN 50020 — Intrinsic safety "i"

In addition to the Common Market countries, other Western European countries adopted these standards as the European Norms. The Common Market eventually evolved into the European Community, and in the 1980s directives were published concerning electrical equipment for use in potentially explosive atmospheres. These require that the member countries incorporate the EN standards into their national standards. In 1994 it issued the ATEX Directive, which mandates that all equipment of this type put into service after June 2003 have the CE mark affixed indicating conformity with the EN standards.

6-1.3 Foreign National Standards

All of the larger industrialized countries have standards organizations similar to ANSI in the United States. These organizations may be part of, or at least partially financed by, the government. The U.S. voluntary system is not financed by the government.

6-1.4 List of Applicable Standards

Table 6-1 shows national standards organizations, their addresses, and most of the applicable standards that apply to hazardous locations. Several of these standards are available only in the language of the nation of origin. The requirements in Canada and the United States are very similar.

6-2 Comparison of International and Foreign Standards

6-2.1 Definitions of Locations, Gases, and Vapors

6-2.1.1 Class

In almost all countries, the mining equipment for use in gassy mines is covered by special standards. In the IEC and many foreign standards,

Table 6-1 National Standards Organizations

Country	Standards Organization	Standard Number	Description
Australia	Standards Association of Australia (SAA) 80-86 Arthur Street North Sydney NSW 2060 Australia	AS1076	Selection, installation, and maintenance of electrical equipment
		Part 1	Basic requirements
		Part 3	Flameproof
		Part 4	Intrinsically safe
		Part 5	Pressurization
		Part 6	Increased safety
		Part 7	Nonsparking
		Part 8	Special protection
		Part 9	Oil immersed and sand filled
		Part 10	Combustible dusts
		AS1021	Purging
		AS1593	Increased safety
		AS1825	Pressurized
		AS1826	Special protection
		AS1828	Cable glands
		AS1829	Intrinsically safe
		AS2236	Dust-excluding ignitionproof
		AS2238	Nonsparking
		AS2380	
		Part 1	General requirements
		AS2430	Classification of areas
		Part 1	Gas atmospheres
		Part 2	Dust
		AS2431	Encapsulated
		AS2480	Flameproof
Austria	Osterreichescher Verband für Electrotechnik (OVE) Eschenbachgasse 9 Wien 1 Austria	OVE 71	Equipment
		OVE Ex 65	Installation
		In addition, see the following addenda:	
		E71a	
		E71b	
		E71c	
		CENELEC Standards	
Belgium	Belgian Standards Institute (IBN) 29 Avenue de la Brabanconne Brussels 4 Belgium	Follows CENELEC	
		C23-001 (EN50014)	General requirements
		C23-101 (EN50020)	Intrinsically safe
		C23-102 (EN50019)	Increased safety
		C23-103 (EN50018)	Flameproof
		C23-104 (EN50015)	Oil immersion
		C23-105 (EN50016)	Pressurized
		C23-106 (EN50017)	Powder filled

Table 6-1 ***(continued)***

Country	Standards Organization	Standard Number	Description
		C23-201 (EN50039)	Intrinsically safe
		RGPT Article 251 bis	Installations
Canada	Canadian Standards Association 178 Rexdale Boulevard Rexdale, Ontario M9W 1R3 Canada	C22.2 No. 174	Cable and glands
		C22.2 No. 25	Enclosures — dust
		C22.2 No. 30	Enclosures — explosion-proof
		C22.2 No. 137	Luminaires
		C22.2 No. 145	Motors
		C22.2 No. 157	Intrinsically safe
		C22.2 No. 159	Plugs and receptacles
		C 22.1	Installation
Denmark	Dansk Standardizeringsraad Aurehojvej 12 2900 Hellerup Copenhagen Denmark	Follows CENELEC	
		AFSNIT 7A	Installation
		AFSNIT 50	General requirements
		AFSNIT 50-1	Oil immersion
		AFSNIT 50-2	Pressurized
		AFSNIT 50-3	Powder filling
		AFSNIT 50-4	Flameproof
		AFSNIT 50-5	Increased safety
		AFSNIT 50-6	Intrinsic safety
Finland	Finnish Standards Association (SFS) Suomen Standardisoimislüto Bulevardi 5. A7 Helsinki Finland	Eliktriska Inspektoratets Issues "Electrical Safety Regulations"	
		Section 41	Installations in explosion-risk areas
		Follows CENELEC	
		SFS4094	Covers all types of protection
		SFS2972	Dusttight
France	Union Technique de l'Electricité 20 Rue Hamelin Paris 16e France	Follows CENELEC	General requirements
		NF C23-514	Oil immersion
		NF C23-515	Pressurized
		NF C23-516	Powder filling
		NF C23-517	Flameproof
		NF C23-518	Increased safety
		NF C23-519	Intrinsically safe,
		NF C23-520	Type "i"
		NF C23-539	Intrinsically safe
Germany	VerbandDeutscher Elektrotechniker e.V. Vorschriftenstelle 6 Frankfurt/Main 70 Stresemannalle 21 Germany	VDEO 165/9.83	Installation
		Follows CENELEC	
		VDEO 170/0171	Covers above-ground equipment

Table 6-1 ***(continued)***

Country	Standards Organization	Standard Number	Description
Greece	General Directorate of Industry — Directorate of Standardization General Directorate II Ministry of Industry Kanengos Street Athens Greece	Follows CENELEC	
Italy	Comitato Electrotechnico Italiano Via San Paolo 10 Milan Italy	Follows CENELEC	
		CEI31-1	Flameproof
		CEI31-2	Pressurized
		CEI31-5	Oil immersion
		CEI31-6	Powder filling
		CEI31-7	Increased safety
		CEI31-8	General requirements
		CEI31-9	Intrinsic safety "i"
		CEI31-10	Intrinsically safe
		CEI31-11	Nonsparking
		CEI64-2	Installations
Japan	Japanese Standards Association 1-24 Akasaka 4-Chome Minato-ku Kitamagun Tokyo Japan	JIS C0903	General rules for electrical apparatus
		JIS C0905	Construction of electrical power apparatus
Netherlands	Nederlands Normalisate Institute Polakweg 5 Rijswukr (2H) Netherlands	Follows CENELEC	
		NEN1010	Wiring regulations
		NEN3125	Hermetically sealed and nonsparking
		NEN EN50014	General requirements
		NEN EN50015	Oil immersed
		NEN EN50016	Pressurized
		NEN EN50017	Powder filled
		NEN EN50018	Flameproof
		NEN EN50019	Increased safety
		NEN EN50020	Intrinsic safety "i"
Norway	Norges Standardiserings Forbund Haakons VII Gate 2 Oslo 1 Norway	Follows CENELEC	
		NEN110.83	
		NEN111.83	General requirements
		NEN112.83	Oil immersed
		NEN113.83	Pressurized
		NEN114.83	Powder filled
		NEN115.83	Flameproof
		NEN116.83	Increased safety
		NVE	Intrinsic safety "i"
		Communication 1/77	Wiring rules

Table 6-1 ***(continued)***

Country	Standards Organization	Standard Number	Description
South Africa	South African Bureau of Standards	SABS086	Installation and maintenance
	Private Bag 191	SABS0108	Classification of areas
	Pretoria	SABS0142	Wiring
	Republic of South Africa	SABS314	Flameproof
		SABS549	Intrinsically safe
		SABS969	Dust-ignitionproof
		SABA970	Nonsparking
		SABA1031	Increased safety
		SABS808	Cable glands

this equipment is placed in what is known as Group I. The equipment addressed here is for use in aboveground installations, such as oil refineries, and petrochemical and chemical plant installations. In the IEC and many countries, this equipment is in Group II. In the United States, it is in Class I.

6-2.1.2 Group

Group II covers all of the flammable gases and vapors and is further subdivided into three subgroups: A, B, and C. Typical examples are shown in Table 6-2. The grouping of gases is based primarily on the maximum experimental safe gap (MESG) and/or minimum ignition current (MIC). This requires special test equipment. The MESG is done in a chamber with a known mixture of flammable gas. One end of the chamber is supplied with a fixed flat flame path [usually 1 in. (25 mm)] with a variable gap. The flammable gas is ignited with a fixed gap. The gap is gradually closed until there is no propagation through the gap. The maximum gap that does not propagate flame is known as the MESG.

For intrinsically safe equipment the minimum ignition current is important. The equipment used is the IEC break-spark apparatus originally developed by the German PTB Laboratory. Many chemicals have been tested and their MIC values determined.

Work done at the Electrical Research Association in the United Kingdom has shown that there is fairly good correlation between the MESG and the MIC for the various chemicals. This report would indicate that if the MIC is known, the assignment of the appropriate group for the gas is possible.

The grouping of gases is similar among the IEC, Europe, and the United States There are a few exceptions, however, since IEC tests do not consider the pressure-piling or the turbulent effects on the MESG. In addition, in the United States, carbon disulfide is not classified, and

Table 6-2 Overseas Grouping of Gases

I (Mining)	IIA	IIB	IIC
Methane	Acetaldehyde Acetic acid Acetone Aircraft fuel Ammonia Benzene Carbon monoxide Diesel Ethane Ethyl acetate Ethanol Heating fuel Vinyl acetate Methanol Butane Butanol Hexane Petrol Propane Toluene Turpentine	Diethyl ether Ethylene Coke oven gas	Acetylene Carbon disulphide Hydrogen

Some examples of gases that are grouped differently are:

	European	U.S.
Acetaldehyde	Group IIA	Class I, Group C
Carbon monoxide	Group IIA	Class I, Group C

hydrogen and acetylene are in separate groups. Table 6-2 shows the overseas grouping practice for a few gases. See Appendix A, Table A-4, for the U.S. grouping.

An example of a gas that is classified higher due to pressure-piling effects is ethylene oxide. In the United States, it is placed in Class I, Group B, but by using seals on all conduit entries, it is possible to use Class I, Group C equipment. (See also Chapter 2).

In addition to the grouping of gases, the maximum surface temperature of the electrical equipment is important due to autoignition of the flammable gas. All countries require electrical equipment to be marked with its temperature class when operating in a 104°F (40°C) ambient. In the U.S. division system, 14 classes are used, while overseas only six are utilized. A comparison of the temperature classes is shown in Table 6-3.

6-2.1.3 Division

In addition to the grouping of the gas and its ignition temperature, it is necessary to determine the division or zone. In order to select the correct

Table 6-3 Comparison of Electrical Equipment Temperature Classes

Europe or IEC		U.S. Division System		
			Maximum Surface Temperature	
Tempeature Class	Maximum Surface Temperature (° C)	Temperature Class	°C	°F
T1	450	T1	450	842
T2	300	T2	300	572
		T2A	280	536
		T2B	260	500
		T2C	230	446
		T2D	215	419
T3	200	T3	200	392
		T3A	180	356
		T3B	165	329
		T3C	160	320
T4	135	T4	135	275
		T4A	120	248
T5	100	T5	100	212
T6	85	T6	85	185

electrical equipment, it is necessary to determine the classification of the location where this equipment is to be located. In the IEC and most overseas countries there are three zones (divisions) where the flammable gas may occur:

Zone 0 — An area that is always hazardous or hazardous for long periods of time, for example, vapor space of a process vessel

Zone 1 — An area in which an explosive mixture is likely to occur during normal operation

Zone 2 — An area in which an explosive mixture is not likely to occur, and if it occurs, will only occur for a short period of time

These zones are equivalent to the two divisions used in the United States. The U.S. concept of Division 1 combines Zones 0 and 1. Zone 2 and Division 2 are similar and can be considered the same. Since Division 1 can be hazardous all of the time, some types of construction used overseas in Zone 1 are not permitted in the U.S. Division 1.

6-2.2 Definitions of Locations, Combustible Dusts

The grouping of combustible dusts is not covered or as well defined in the IEC or other national standards as it is in the United States and Canada. However, the IEC is presently considering the use of resistivity of combustible dusts as one of the criteria for classifying dusts. Table 6-4 shows the countries that currently have information on combustible dusts in their standards.

Table 6-4 Combustible Dusts

Country	Classification of Areas	Groupings of Dusts and Standards	Examples
Australia	Classification of Areas — SAA Wiring Rules: Clause 9.2 and Appendix F. Class II Hazardous Dusts, Fibers, or Flyings — Areas that are hazardous because of the presence of combustible dust, fibers, or flyings. Class II, Division 1 — Areas in which: (i) combustible dusts, fibers, or flyings of an electrically conductive nature are present regardless of particle size; or (ii) electrically nonconductive dusts, fibers, or flyings of such fineness as to be capable of producing explosive mixtures when suspended in air are present not in suspension, but lying as settled dust; or which may be in suspension either continuously, intermittently, or periodically under normal operating conditions in quantities sufficient to produce an explosive concentration; or which are present in accumulations that may be thrown into suspension in air in sufficient quantities to produce an explosive concentration, by mechanical failure or abnormal operation of plant. Class II, Division 2 — Areas in which: (i) electrically non-conductive combustible dusts, or fibers or flyings, of such coarseness as to be incapable of remaining in suspension in air in quantities sufficient to produce an explosive concentration are present, but where accumulation of such substances may be sufficient to interfere with the safe dissipation of heat from electrical equipment; or (ii) deposits of	Class A — Inherently explosive dusts regardless of particle size are liable to explode at temperatures of 100°C. Class B1 — Inherently explosive dusts regardless of particle size that are dangerous when exposed to temperatures above 100°C. Class B2 — All electrically conductive dusts regardless of particle size. Class B3 — All nonconductive combustible dusts that have a settling rate of or slower than 5 mm/second in still air. Class C1 — All nonconductive combustible dusts that have a settling rate faster than 5 mm/second in still air. Class C2 — All dusts that when settled on plant surfaces may be susceptible to spontaneous combustion or easy ignition due to carbonization by excessive dryness resulting from exposure to heat of the plant surfaces. Class C3 — All dusts that when settled on surfaces of plant enclosures have an insulating effect that would result in the overheating, malfunction, or deterioration of the enclosed plant and decrease safety in operation.	Grouping of typical hazardous dusts: Group A: Nitro starch Lead azide Lead styphnate Group B1: Gunpowder Guncotton dry Propellant powder Tetryl dust Tetryl 100 mesh Group B2: Graphite Coal Coke Charcoal Coal tar pitch Metal powders Group B3: Phenolic resin Polystyrene Shellac, resin, gum Vinyl butyral resin Clover seed Garlic dehydrated Rice Sulphur Group C: Rayon Cotton lintels Cotton waste Sisal Heniquin Istle Jute Hemp Tow Cocoa fiber Oakum Spanish moss Excelsior Peanut cracklings

Table 6-4 ***(continued)***

Country	Classification of Areas	Groupings of Dusts and Standards	Examples
	such material may become susceptible to spontaneous combustion or easy ignition due to carbonization, or excessive dryness, resulting from exposure to heat dissipated from electrical equipment. AS2430, Part 2 shows Zones 3, 4, and 5, which do not agree with the following, which shows this method of classification: Group A Dusts — The zonal classifications for Group A dusts shall be as follows: (a) Where a Group A dust is exposed in any quantity in an enclosed space, the entire space shall be Zone 3. (b) Where the enclosed space has an opening to free space, the space adjoining the opening equal to twice the height of the opening vertically and twice the width horizontally in either direction shall be Zone 3. (c) Where the enclosed space has an opening to an adjoining enclosed space, the adjoining space shall be Zone 3. Group B Dusts — The zonal classification for Group B dusts shall be as follows: (a) Group B1 and B2 dusts: (i) Where a Group B1 and B2 dust is exposed in any quantity in an enclosed space, the whole space shall be Zone 4. (ii) Where the enclosed space has an opening to free space, the space adjoining the opening equal to twice the height of the opening vertically and twice the width horizontally in either direction shall be Zone 5. (iii) Where the enclosed space has an opening to an adjoining enclosed		

Table 6-4 ***(continued)***

Country	Classification of Areas	Groupings of Dusts and Standards	Examples
	space, the adjoining space shall be Zone 5. (iv) Free space adjoining an opening shall be nonhazardous. Group C Dust — Where a Group C dust is exposed in quantities that constitute a hazard in an enclosed space, the areas of settlement of the dust shall be Zone 5. Sealed Containers — Areas in which Groups A and B dusts are stored in sealed containers and remain sealed shall be Zone 5. Outdoor Situations — Open space containing dust is nonhazardous in the normal circumstances. However, when such conditions as still air, very slow dispersion of dust cloud, or any other situation in which a likely dust hazard could arise, the space concerned shall be Zone 5.		
Denmark	Combustible Dusts 10 — Area in which a potentially explosive atmosphere in the form of a flammable dust can occur often or for long periods. Combustible Dusts 11 — Area in which a potentially explosive atmosphere can occur for short periods due to the disturbance of flammable dusts.	AFSNIT 7A Appendix 1	
Finland	Flammable Dusts — Locations where flammable dusts in explosive concentrations are present in the air.	Electrical Safety Regulations Appendix V	
Germany	Zone 10 — Area in which an explosive concentration of dust is present continuously or for a long period.		

Table 6-4 ***(continued)***

Country	Classification of Areas	Groupings of Dusts and Standards	Examples
	Zone 11 — Area in which an explosive concentration of dust may be present occasionally for a short period as a result of the disturbance of dust deposits.		
Italy	Zone Class II — Location in which hazardous dusts are present. Zone Class III — Location in which hazardous combustibles such as fibers are present. It also covers locations with gases or liquids having either flash points above 40°C if not processed or stored in such a way as to allow contact with air at ambients above flash point, or in quantities less than those prescribed for a Class I location.		
Japan	Flammable Dusts — Dusts that burn with the occurrence of an exothermic reaction using the oxygen present in air. 1. Electrical nonconductive flammable dust — Dusts such as wheat flour, starch, sugar, and synthetic resin. The dust of flammable fibers such as cotton, silk, and rayon are included, but the fibers themselves are excluded. 2. Electrically conductive flammable dust — Dusts such as carbon black, coke, iron, and copper. Explosive Dusts — Metallic dust that burns even in rarefied oxygen atmospheres or carbon dioxide and that will explode when in the suspended state. Dusts such as magnesium and aluminum are included. (Substances and mixtures such as explosives that have intromolecular oxygen, or that	RIIS — Recommended Practice for Electrical Equipment for Use in Explosive Dust Atmospheres in General Industries	

Table 6-4 ***(continued)***

Country	Classification of Areas	Groupings of Dusts and Standards	Examples
	are mixed with oxides and do not require oxygen to burn, are excluded.) Explosive Dust Locations — Equipment to be used: special dustproof. Flammable Dust Locations — 1. Electrically nonconductive dust — Equipment to be used: dustproof. 2. Electrically conductive dust — Equipment to be used: special dustproof.		
Norway	Category 'b' — Location where the risk of explosion is due to air mixed with a flammable dust. Category 'b,' Zone 0 — Location in which a considerable quantity of flammable dust, formed as explosive dust clouds in the air, is continuously present or present for a long period. Category 'b,' Zone 1 — Location in which a considerable quantity of flammable dust, formed as explosive dust clouds in the air, is likely to occur in normal operation. Category 'c' — Location where the risk of explosion is due to the presence of an explosive substance. Category 'd' — Location where the risk of explosion is due to a mixture of flammable gas, vapor or suspended liquids, or dust together with an atmosphere containing greater amounts of oxygen than present in normal air, or with other gases, which more readily support combustion than normal air.	NYE Communication 1/77 Appendix III	
South Africa	Similar to USA		

Table 6-4 ***(continued)***

Country	Classification of Areas	Groupings of Dusts and Standards	Examples
Spain	Class II — Locations that are hazardous because of the presence of combustible dust. Class II, Division 1 — Locations: (1) in which combustible dust exists or may be in suspension in the air continuously, intermittently, or periodically, under normal operating conditions, and is in quantities sufficient to produce explosive or ignitible mixtures; (2) in which mechanical failure or abnormal operation of machinery or equipment might cause such mixtures to be produced and might also provide a source of ignition through simultaneous failure of electrical equipment, operation of protection devices, or from other causes; (3) in which dusts of an electrically conducting nature may be present. Class II, Division 2 — Locations in which combustible dust will not normally be in suspension in the air or will not be likely to be thrown into suspension by the normal operation of equipment or apparatus, in quantities sufficient to produce explosive or ignitible mixtures but: (1) where deposits or accumulations of such dust may be sufficient to interfere with the safe dissipation of heat from electrical equipment or apparatus; (2) where such deposits or accumulations on, inside, or in the vicinity of electrical equipment might be ignited by arcs, sparks, or burning material from such equipment; (3) that are adjacent to Class II, Division 1 loca-		

Table 6-4 ***(continued)***

Country	Classification of Areas	Groupings of Dusts and Standards	Examples
	tions from which hazardous concentrations of dust in suspension may be communicated in conditions of abnormal operation.		
	Class III — Locations that are hazardous because of the presence of easily ignitible fibers or flyings but in which such fibers or flyings are not likely to be in suspension in the air in quantities sufficient to produce ignitible mixtures.		
	Class III, Division 1 — Locations in which easily ignitible fibers, or materials producing combustible flyings, are handled, manufactured, or used.		
	Class III, Division 2 — Locations in which easily ignitible fibers are stored or handled (except in process of manufacture).		

The installation rules of each country listed in Table 6-4 determine the types of protection that may be used in the various zones. In addition, they specify the type of wiring systems that are permitted. This is similar to the requirements in Articles 500 through 503 of the *National Electrical Code*[2].

6-2.3 Types of Protection Permitted and Where They May Be Used, as Compared to the United States

There are many types of protection that have been recognized as suitable for hazardous locations overseas. Not all of them are permitted in the United States division system. Some of the common types of protection recognized in the U.S. are explosionproof, intrinsically safe, purged and pressurized, and oil immersed. Table 6-5 shows the various types of protection and their definitions. The third column shows the IEC symbol. Since CENELEC and some foreign standards have adopted a modified IEC standard, these symbols are used overseas in many countries. The fourth column shows the zone where these types may be used. The division reference is for the United States.

The various national installation standards also modify the definitions of the various zones. Table 6-6 lists the countries and their appro-

Table 6-5 Protection Types

Type of Protection	Definition	IEC and CENELEC Symbol	Where Used
Explosionproof (Flameproof)	An enclosure for electrical equipment that will withstand, without damage, an explosion of a prescribed flammable gas or vapor within the enclosure and prevent the transmission of flame or sparks such as would ignite the external prescribed flammable gas or vapor for which it is designed, and that normally operates at an external temperature that will not ignite the external prescribed flammable gas or vapor. A flameproof enclosure will not necessarily or ordinarily be weatherproof or dustproof.	Ex d	Divisions 1 and 2 Zones 1 and 2
Pressurized	An enclosure for electrical machines and equipment in which the entry of flammable gases or vapors is prevented by maintaining the air (or other nonflammable gas) within the enclosure at a specified pressure above that of the external atmosphere.	Ex p	Divisions 1 and 2 Zones 1 and 2
Purged	An enclosure for electrical equipment in which a sufficient flow of fresh air or inert gas is maintained through the enclosure to prevent the entry of any flammable gas or vapor that may be present in the ambient atmosphere in which the enclosure is installed.	Ex p	Divisions 1 and 2 Zones 1 and 2
Oil Immersed	Electrical equipment of which all parts on which arcs may occur in normal service are immersed in oil to a sufficient depth to prevent ignition of an explosive gas/air mixture that may be present above the surface of the oil, and all live parts on which arcs do not occur in normal service are either immersed in oil or protected by some other recognized technique.	Ex o	Divisions 1 and 2 Zones 1 and 2
Powder Filled	An enclosure for electrical equipment in which all live parts of the equipment are entirely embedded in a mass of powdery material having the consistency of sand in such a way that if, under the conditions of use for which the equipment is designed, an arc occurs within the enclosure, this arc will not ignite the	Ex q	Zones 1 and 2

Table 6-5 ***(continued)***

Type of Protection	Definition	IEC and CENELEC Symbol	Where Used
	outer explosive atmosphere either by the transmission of flame or by the overheating of the walls of the enclosure.		
Increased Safety	A method of protection in which measures additional to those adopted in ordinary industrial practice are applied, so as to give increased security against the possibility of excessive temperatures and the occurrence of arcs or sparks in electrical apparatus that does not produce arcs or sparks in normal service.	Ex e	Zones 1 and 2
Intrinsically Safe	A circuit or part of a circuit is intrinsically safe when any sparking that is produced normally by breaking or making the circuit or produced accidentally (that is, by short-circuit or earth-fault) is incapable under prescribed test conditions of causing ignition of a prescribed gas or vapor.	Ex i_a Ex i_b	Divisions 1 and 2 Zones 0, 1, and 2 Division 2 Zone 2
Special Protection	A concept that has been adopted to permit the certification of those types of electrical apparatus that, by their nature, do not comply with the constructional or other requirements specified for apparatus with established types of protection but that nevertheless can be shown, where necessary by test, to be suitable for use in hazardous areas, in prescribed zones.	Ex s	Depending on the qualification on the certificate
Nonsparking	Electrical equipment of normal industrial design having no sparking parts or hot spots in normal operation.	Ex n	Zone 2
Encapsulation	The intimate surrounding of electrical equipment that has been treated with a suitable material in such a way that under the conditions of use for which it is designed, no external condition can occur that will ignite the outer explosive atmosphere either by transmission of flame or overheating of any part.	Ex m	Zones 1 and 2

Table 6-6 Information by Country on Classification and Protection Systems

Country	Grouping of Gases	Classification of Areas	Construction Permitted	Installation Rules	Electrical Supply
Australia	Explosion characteristics ignition temperature MESG or MIC	Zone 0 — Areas where explosive gas/air mixtures may be continuously present or present for long periods. Zone 1 — Areas where explosive gas/air mixtures exist intermittently or periodically under normal operating conditions or may exist frequently due to leakage. Zone 2 — Areas where explosive gas/air mixtures are not likely to occur and if they do occur will exist only for a short time.	Flameproof Pressurized Purged Ventilation Oil immersed Powder filled Increased safety Intinsically safe Special protection Nonsparking Encapsulation	AS3000-1981	415/240V 3 Phase 4W with neutral earthed 50 Hz; West Australia 440/250 V 3 Phase 4W 50 Hz, with neutral earthed
Austria	Follows CENELEC	Follows CENELEC	Follows CENELEC	OVE-EX65	Normally 380/220 V 3 Phase 4W 50 Hz, and neutral normally earthed
Belgium	Follows CENELEC	Follows CENELEC	Follows CENELEC	Article 215 bix of RGPT	380 V 3 Phase 4W 50 Hz and some 500 V 3 Phase 50 Hz, neutral may be earthed.
Denmark	Follows CENELEC	Follows CENELEC	Follows CENELEC	AFSNIT7A	380 V 3 Phase 4W 50 Hz, with neutral earthed
Finland	Follows CENELEC	Follows CENELEC	Follows CENELEC	Electrical Safety Regulations Sections 37, 38, and 41	380/220 V 3 Phase 4W 50 Hz, with neutral earthed
France	Follows CENELEC	Follows CENELEC	Follows CENELEC	Decret No. 62-1454 Journal Officiel No. 1078 UTE C12-100	380 V 3 Phase 4W 50 Hz and some 220 V 3 Phase 4W 50 Hz, neutral normally earthed

Table 6-6 *(continued)*

Country	Grouping of Gases	Classification of Areas	Construction Permitted	Installation Rules	Electrical Supply
Germany	Follows CENELEC	Follows CENELEC	Follows CENELEC	VDEO107 VDEO165 Elexv EX-RL	380/220 V3 Phase 4W 50 Hz, with neutral earthed
Greece	Follows CENELEC	Follows CENELEC	Follows CENELEC	Regulations for Interior Electric Installations Chapter X (Government Gazette No. 735B-12/9/66)	380/220 V 3 Phase 4W 50 Hz
Italy	Follows CENELEC	Follows CENELEC	Follows CENELEC	CE164-2	380/220 V 3 Phase 4W 50 Hz, with neutral earthed
Japan	Similar to German practice	Divisions 0, 1, and 2 are similar to Australian Zones 0, 1, and 2	Flameproof Pressurized Oil immersed Sand filled Increased safety Intrinsically safe Special protection Plate protected	RIIS-Recommended Practice for Explosion-Protected Electrical Installations in General Industries	200/100 1 Phase 50 or 60 Hz, with center tapped 200 V 3 Phase 3W delta 50 or 60 Hz
Netherlands	Follows CENELEC	Follows CENELEC	Follows CENELEC	NEN1010 V1041	380/220 V 3 Phase 4W 50 Hz and some 500 V 30 50 Hz, neutral is normally earthed
Norway	Follows CENELEC	Follows CENELEC	Follows CENELEC	NVE Communication No. 1/77	230 V 3 Phase 3W 50 Hz star-connected with unearthed neutral. In addition, there may be 380/220 V3 Phase 50 Hz, 400/

Table 6-6 *(continued)*

Country	Grouping of Gases	Classification of Areas	Construction Permitted	Installation Rules	Electrical Supply
					230 V, 3 Phase 4W 50 Hz both with earthed neutral and 380/400 V 3 Phase 3W 50 Hz, with no earthed neutral.
Portugal		Follows CENELEC	Follows CENELEC	Decree No. 740/74	380/220 V 3 Phase 4W 50 Hz, with neutral earthed
South Africa	According to the explosion characteristics, but hydrogen, ethyl nitrate, and carbon disulfide are not classified.	Class I — Locations in which flammable gases or vapors are or may be present in sufficient quantities to become hazardous. Division 0 — Locations in which flammable gases or vapors are present continuously in concentration between UEL and LEL. Division 1 — Locations in which: (a) hazardous concentrations of flammable gas or vapor occur intermittently or periodically under normal operating conditions, or (b) hazardous concentrations of flammable gas or vapor may occur frequently because of repair or maintenance operations or leakage, or (c) breakdown or faulty operation of equipment or processes, which might release dangerous concentrations of flammable gases or vapors, might also cause simultaneous failure of electrical equipment. Division 1A — In which concentrations of flammable gases or vapors occur inter- mittently or periodically under normal operating conditions, or in which concentrations of flammable gases or vapors may occur frequently due to repair or maintenance operations or leakage, but in which operating conditions and the nature of the flammable	Flameproof Increased Safety Intrinsically Safe Nonsparking	SABS086	380 V 3 Phase 4W 50, Hz, with neutral earthed and some 430 V 3 Phase 4W 50 Hz

Table 6-6 ***(continued)***

Country	Grouping of Gases	Classification of Areas	Construction Permitted	Installation Rules	Electrical Supply
		gases or vapors released are such that safety measures against the occurrence of explosive concentrations are taken for reasons additional to the explosion hazard. Division 2—Locations in which operations concerned with flammable or explosive substances, gases or vapors, or volatile liquids are so well controlled that an explosive or ignitible concentration is only likely to occur under abnormal conditions. SABS0108 also recognizes the *NEC*, API 500 Series, and the U.K. Institute of Petroleum, *Model Code of Safe Practice in the Petroleum Industry.*			
Spain	Follows CENELEC	Follows CENELEC	Follows CENELEC	Instruccion MI.BT026	220/127 V3 Phase 4W 50 Hz or 380/220 V 3 Phase 4W, both with neutral earthed
Sweden	Follows CENELEC	Follows CENELEC	Follows CENELEC	Safety Regulations, Sections 57 to 59 SS4210821	380 V 3 Phase 4W 50 Hz, with neutral earthed. Processing industries may have 380 V or 3 Phase 50 Hz isolated.
Switzerland	Follows CENELEC	Follows CENELEC	Follows CENELEC	SEV1000, but does not have a specific section on hazardous locations	380/220 V 3 Phase 4W 50 Hz and 500 V 3 Phase 3W 50 Hz
U.K.	Follows CENELEC	Similar to Australia Follows CENELEC	Follows CENELC	BS5345	415/240 V3 Phase 4W 50 Hz, with neutral normally earthed

priate definitions, how the gases are grouped, types of protection permitted, the installation standard, and the electrical supply.

Explosionproof or flameproof enclosures have some similarities. All countries require a rugged housing that will withstand an internal explosion of a typical gas, will not permit hot sparks or gases to escape, and will operate at a temperature that will not ignite the surrounding hazardous atmosphere. The tests used to determine these features are similar, except in the United States these enclosures would be tested with various lengths of conduit. This additional testing considers the pressure-piling effect and determines the maximum pressure developed by the internal explosion.

The marking on the nameplate would also be different. In the U.S. division system, an explosionproof switch would be marked as Class I, Group D. Other countries would require ''Ex d IIA T6'' to identify a flameproof device suitable for hydrocarbons that operates at less than 185°F (85°C). In the United States the temperature class is not required when the maximum surface temperature is less than 212°F (100°C).

In some countries, all flameproof enclosures must be routinely tested hydrostatically at a pressure of $1^1/_2$ to 2 times the maximum explosion pressure. In the United States, the enclosure is tested at four times the explosion pressure when it is listed or certified. This eliminates routine testing.

For combustible dust locations, many countries specify enclosures that exclude dust and have surface temperatures less than the ignition temperature of the dust. In the United States, these are known as dust-ignitionproof enclosures, while other countries use the term ''dusttight.''

The IEC Standard 60529 is used by most overseas countries to determine if an enclosure will exclude dust. For conductive dusts, the enclosure must pass the IP 65 tests; for nonconductive dusts, the enclosure must pass the IP 54 tests. The main difference is that IP 65 does not permit any entrance of dust, while IP 54 will permit the entrance of a limited amount of dust if it does not interfere with the operation of the interior equipment. The first digit refers to the ingress of foreign solid bodies, while the second digit refers to ingress of water. In all cases, the higher the number, the more severe is the test. Tables 6-7 and 6-8 provide more information on the numbering system.

In the United States, a similar enclosure designation is the National Electrical Manufacturers Association (NEMA) standard. It has recently been included in the *NEC* and is part of the Underwriters Laboratories Inc. (UL) standards. The document includes other tests, such as corrosion and oil spray, which are not in IEC Standard 529. Table 6-9 shows the conversion from NEMA types to the IP system. Since the tests are different, the conversion is based on a very conservative approach.

Table 6-7 Degrees of Protection Indicated by the First Characteristic Numeral

First Characteristic Numeral	Degree of Protection	
	Short Description	Definition
0	Nonprotected	No special protection
1	Protected against solid objects greater than 50 mm (1.97 in.)	A large surface of the body, such as a hand (but no protection against deliberate access). Solid objects exceeding 50 mm (1.97 in.) in diameter
2	Protected against solid objects greater than 12 mm (0.47 in.)	Fingers or similar objects not exceeding 80 mm (3.15 in.) in diameter
3	Protected against solid objects greater than 2.5 mm (0.098 in.)	Tools, wires, and so forth, of diameter or thickness greater than 2.5 mm (0.098 in.). Solid objects exceeding 2.5 mm (0.098 in.) in diameter.
4	Protected against solid objects greater than 1.0 mm (0.039 in.)	Wires or strips of thickness greater than 1.0 mm (0.039 in.). Solid objects exceeding 1.0 mm (0.039 in.) in diameter.
5	Dust-protected	Ingress of dust is not totally prevented but dust does not enter in sufficient quantity to interfere with satisfactory operation of the equipment.
6	Dusttight	No ingress of dust

6-2.4 Listing or Certification Requirements

In all major countries, certification or listing of electrical equipment in hazardous locations is mandatory. Most countries require the identification mark of the approvals laboratory on the product. In addition, many countries outside of North America issue a certificate of conformity to the manufacturer. This document lists the various conditions of use for the device. In some cases, the users require a copy of the certificate for their records.

One of the goals of CENELEC when they harmonized the standards for hazardous locations was to permit the free flow of these products among the member countries. The CENELEC directive implies reciprocity of certification between the approval organizations. This goal has not been completely reached, since some users and inspection authorities insist on certification by their own national authority. The IEC Ex scheme as described in Section 6-1.1, however, is a further effort towards this goal.

Table 6-8 Degrees of Protection Indicated by the Second Characteristic Numeral

Second Characteristic Numeral	Degree of Protection	
	Short Description	Definition
0	Nonprotected	No special protection
1	Protected against dripping water	Dripping water (vertically falling drops) shall have no harmful effect.
2	Protected against dripping water when tilted up to 15 degrees	Vertically dripping water shall have no harmful effect when the enclosure is tilted at any angle up to 15 degrees from its normal position.
3	Protected against spraying water	Water falling as a spray at an angle up to 60 degrees from the vertical shall have no harmful effect.
4	Protected against splashing water	Water splashed against the enclosure from any direction shall have no harmful effect.
5	Protected against water jets	Water from heavy seas or water projected in powerful jets shall not enter the enclosure in harmful quantities.
6	Protected against heavy seas	Water from heavy seas or water projected in powerful jets shall not enter the enclosure in harmful quantities.
7	Protected against the effects of immersion	Ingress of water in a harmful quantity shall not be possible when the enclosure is immersed in water under defined conditions of pressure and time.
8	Protected against submersion	The equipment is suitable for continuous submersion in water under conditions that shall be specified by the manufacturer. Note: Normally, this will mean that the equipment is hermetically sealed. However, with certain types of equipment, it can mean that water can enter, but only in such a manner that it produces no harmful effects.

6-2.4.1 Testing Authorities

Some of the major testing authorities and their addresses are shown as follows:

Canada

Canadian Standards Association
178 Rexdale Boulevard
Etobicoke, Ontario M9W 1R3
Canada

Table 6-9 Conversion of NEMA-Type Numbers to IEC Classification Designations*

NEMA Enclosure Type Number	IEC Enclosure Classification
1	IP 10
2	IP 11
3	IP 54
3R	IP 14
3S	IP 54
4 and 4X	IP 56
5	IP 52
6 and 6P	IP 67
12 and 12K	IP 52
13	IP 54

**Cannot be used to convert IEC classification designations to NEMA-type numbers.*

France

Laboratoire Central des Industries Electriques (LCIE)
333 Avenue du Général Leclerc
92260 Fontenay-aux Roses
France

Cerchar
Laboratories de Verneuil-en-Hallatte
Boite Postale No. 27
60-Creil
France

Germany

Physikalisch-Technische Bundesanstalt (PTB)
Bundesallee 100
D-3300 Braunschweig
Germany

Italy

Centro Elettrotecnico Sperimentale Italiano (CESI)
Via Rubittano 54
20134 Milano
Italy

Norway

Norges Elektriske Materiellkontroll (NEMKO)
Gaustadallen 30
Postboks 288
Blindern
Oslo 3
Norway

United Kingdom

British Approvals Service for Electrical Equipment in Flammable Atmospheres (BASEEFA)
Harpur Hill
Buxton
Derbyshire SK17 9JN
United Kingdom

Many standards for electrical equipment in hazardous locations are in use around the world. The CENELEC harmonization standards have eliminated some differences in construction in Western Europe. The installation rules and classification of areas have several differences. All of the standards are closer today than they were 30 years ago, but it is doubtful that all of the requirements will ever be completely harmonized. All of the standards-writing organizations are attempting to make compromises, but they also want to maintain the high level of safety that has been exhibited in the past.

Bibliography

[1]IEC 60529, *Classification of Degrees of Protection Provided by Enclosures*, International Electrotechnical Commission, Geneva, Switzerland, 1976.

[2]NFPA 70, *National Electrical Code*, National Fire Protection Association, Quincy, MA, 1996.

Tables

Table A-1 Flashpoint of Common Materials*

Material	Flashpoint °F	Flashpoint °C	Material	Flashpoint °F	Flashpoint °C
Acetaldehyde	−38	−39	Isobutyl alcohol	82	28
Acetic acid	103	39	Isopropyl acetate	35	2
Acetone	−4	−20	Isopropyl alcohol	53	12
Acrolein (inhib.)	−15	−26	Jet fuel (A and A-1)	110–150	43–66
Allyl alcohol	70	21	Jet fuel (B and JP-4)	−10 to +30	−23 to −1
Benzene	12	−11	Jet fuel (JP-5)	95–145	35–63
Butane	−76	−60	Jet fuel (JP-6)	100	38
Butyl alcohol	98	37	Kerosene	100–162	38–72
Butylamine	10	−12	Methanol	52	11
Butyl mercaptan	35	2	Methyl acetate	14	−10
Camphor	150	66	Methyl acrylate	27	−3
Carbon disulfide	−22	−30	Methyl ethyl ketone	16	−9
Diethylamine	−9	−23	Methyl isobutyl ketone	64	18
Ethanol	55	13	Methyl methacrylate	50	10
Ethyl acetate	24	−4	Nitromethane	95	35
Ethylbenzene	70	21	Octane	56	13
Ethyl chloride	−58	−50	1-Pentanol	91	33
Ethylene glycol monoethyl ether	110	43	Propylene oxide	−35	−37
Ethyl ether	−49	−45	Styrene	88	31
Formic acid (90%)	122	50	Sulfer	405	207
Fuel oils	100–270	38–132	Toluene	40	4
Gasoline	−45	−43	Turpentine	95	35
Heptane	25	−4	Vinyl acetate	18	−8
Hexane	−7	−22			
Hydrazine	100	38			

**Source: NFPA 325, Guide to Fire Hazard Properties of Flammable Liquids, Gases, and Volatile Solids, National Fire Protection Association, Quincy, MA, 1994.*

Table A-2 Flammable (Explosive) Limits of Common Materials*

Material	Flammable Limits % by Volume	
	Lower	Upper
Acetaldehyde	4	60
Acetic acid	4	19.9
Acetone	2.5	12.8
Acetylene	2.5	100
Acrolein (inhib.)	2.8	31
Allyl alcohol	2.5	18
Allyl ether	0.8	3.4
Ammonia	15	28
Benzene	1.2	7.8
Butane	1.9	8.5
Butyl alcohol	1.4	11.2
Butylamine	1.7	9.8
Butylene	1.6	10
Camphor	0.6	3.5
Carbon disulfide	1.3	50
Carbon monoxide	12.5	74
Diethylamine	1.8	10.1
Ethane	3	12.5
Ethanol	3.3	19
Ethyl acetate	2	11.5
Ethylamine	3.5	14
Ethylbenzene	0.8	6.7
Ethyl chloride	3.8	15.4
Ethylene	2.7	36
Ethylene glycol monoethyl ether	1.7	5.6
Ethylene oxide	3	100
Ethyl ether	1.9	36
Ethyl mercaptan	2.8	18
Formaldehyde (gas)	7	73
Formic acid (90%)	18	57
Fuel Oil No. 1	0.7	5
Gasoline	1.4	7.6
Heptane	1.05	6.7
Hexane	1.1	7.5
Hydrazine	2.9	98
Hydrogen	4	75
Hydrogen sulfide	4	44
Isobutyl alcohol	1.7	10.6
Isopropyl alcohol	2	12.7
Jet fuel (JP-4)	1.3	8
Jet fuel (JP-6)	0.6	3.7
Kerosene	0.7	5
Methane	5	15
Methanol	6	36
Methyl acetate	3.1	16
Methyl acrylate	2.8	25
Methyl ethyl ketone	1.4	11.4
Methyl isobutyl ketone	1.2	8
Methyl methacrylate	1.7	8.2
Naphtha (petroleum)	1.1	5.9
Octane	1	6.5
Pentane	1.5	7.8
1-Pentanol	1.2	10
Propane	2.1	9.5
Propylene	2	11.1
Propylene oxide	2.3	36
Styrene	0.9	6.8
Toluene	1.1	7.1
Vinyl acetate	2.6	13.4

**Source: NFPA 325, Guide to Fire Hazard Properties of Flammable Liquids, Gases, and Volatile Solids, National Fire Protection Association, Quincy, MA, 1994.*

Table A-3 Minimum Explosion Concentration for Some Dusts*

Material	Minimum Explosion Concentration (oz/ft³)
Metal Dusts	
Aluminum, atomized collector fines	0.045
Chrominum (97%), electrolytic, milled	0.230
Ferromanganese, medium carbon	0.130
Iron (98%), H_2 reduced	0.120
Tin (96%), atomized (2% Pb)	0.080
Titanium (99%)	0.045
Vanadium (86.4%)	0.220
Zirconium hydride (93.6% Zr, 2.1% H_2)	0.085
Carbonaceous Dusts	
Charcoal	0.140
Coal, Kentucky bituminous	0.050
Coal, Pittsburgh experimental	0.055
Pitch, coal tar	0.045
Agricultural Dusts	
Alfalfa meal	0.100
Almond shell	0.065
Apricot pit	0.035
Cinnamon	0.06
Cocoa bean shell	0.04
Coconut shell	0.035
Corn	0.055
Cornstarch, commercial	0.045
Cottonseed meal	0.05
Flax shive	0.08
Gum, arabic	0.06
Hemp hurd	0.04
Lycopodium	0.025
Malt barley	0.55
Milk, skimmed	0.05
Peanut hull	0.045
Pectin	0.075
Potato starch dextrinated	0.045
Rice	0.05
Safflower meal	0.055
Soy flour	0.06
Sugar, powdered	0.045
Wheat	0.065
Wheat flour	0.05
Wood flour	0.035
Yeast torula	0.050
Chemicals	
Acetoacetanilide	0.030
Adipic acid	0.035
Benzoic acid	0.030
Dicyclopentadiene dioxide	0.015
Hydroxyethyl cellulose	0.025
Nitrosoamine	0.025
Sorbic acid	0.020
Stearic acid, aluminum salt	0.015
Sulfur	0.035
Drugs	
Aspirin	0.050
Vitamin B1, mononitrate	0.035
Vitamin C (ascorbic acid)	0.070
Dyes, Pigments, Intermediates	
Green base harmon dye	0.030
Red dye intermediate	0.055
Pesticides	
Benzethonium chloride	0.020
Bis (2-hydroxy-5-chlorophenyl) methane	0.040
Crag No. 974	0.025
Ferbam	0.055
Sevin	0.020
Thermoplastic Resins and Molding Compounds	
Acetil Resins	
Acetal, Linear (Polyformaldehyde)	0.035
Acrylic Resins	
Acrylamide polymer	0.040
Acrylonitrile polymer	0.025
Acrylonitrile — vinyl pyridine copolymer	0.020
Methyl methacrylate — ethyl acrylate copolymer	0.030
Methyl methacrylate — ethyl acrylate styrene copolymer	0.025

Table A-3 *(continued)*

Material	Minimum Explosion Concentration (oz/ft³)	Material	Minimum Explosion Concentration (oz/ft³)
Methyl methacrylate — polymer	0.02	Phenol furfural	0.025
Cellulosic Resins		*Phenolic Resins*	
Cellulose acetate	0.035	Phenol formaldehyde	0.025
Cellulose acetate butyrate	0.025	Phenol formaldehyde molding compound (wood flour filler)	0.030
Ethyl cellulose	0.025	Phenol formaldehyde, Polyalkylene-Polyamine modified	0.020
Polyethylene resins		*Polyester Resins*	
Polyethylene, high-pressure process	0.020	Polyethylene terephthalate	0.040
Polyethylene, low-pressure process	0.020	Styrene modified polyester glass fiber mixture	0.045
Polyethylene wax	0.020	*Polyurethane Resins*	
Styrene Resins		Polyurethane foam, fire retardant	0.025
Polystyrene latex	0.020	Polyurethane foam, no fire retardant	0.030
Polystyrene molding compound	0.015	**Special Resins and Molding Compounds**	
Styrene-acrylonitrile (70–30)	0.035	Cashew oil, phenolic, hard	0.025
Vinyl Resins		Lignin, hydrolized, wood-type, fines	0.040
Polyvinyl acetate	0.040	Petrin acrylate monomer	0.045
Polyvinyl butyral	0.020	Petroleum resin (blown asphalt)	0.015
Vinyl-toluene-acrylonitrile Butadiene copolymer	0.020	Rosin, DK	0.015
Thermosetting Resins and Molding Compounds		Rubber, crude, hard	0.025
Allyl Resins		Rubber, synthetic, hard (33% S)	0.030
Allyl alcohol derivative (CR-39)	0.035	Shellac	0.020
Epoxy Resins		Styrene maleic anhydride copolymer	0.030
Epoxy	0.020		
Epoxy — bisphenol A	0.030		

**Source NMAB 353-4, Classification of Dusts Relative to Electrical Equipment in Class II Hazardous Locations, National Academy Press, Washington, DC, 1982. (Available from National Technical Information Service, Springfield, VA 22151.)*

Table A-4 Group Classification of Common Materials*

Material	Group	Material	Group
Acetaldehyde	C	Fuel oils	D
Acetic acid	D	Gasoline	D
Acetone	D	Heptane	D
Acetylene	A	Hexane	D
Acrolein (inhib.)	B(C)+	Hydrazine	C
Allyl alcohol	C	Hydrogen	B
Ammonia	D*	Hydrogen sulfide	C
Benzene	D	Kerosene	D
Butane	D	Methane	D
Butylamine	D	Methanol	D
Butylene	D	Methyl acetate	D
Carbon monoxide	C	Methyl acrylate	D
Diethylamine	C	Methyl ethyl ketone	D
Ethane	D	Methyl isobutyl ketone	D
Ethanol	D	Methyl methacrylate	D
Ethyl acetate	D	Naphtha (petroleum)	D
Ethylamine	D	Nitromethane	C
Ethyl benzene	D	Octane	D
Ethyl butanol	D	Pentane	D
Ethyl chloride	D	1-Pentanol	D
Ethylene	C	Propane	D
Ethylene glycol monoethyl ether	C	Propylene	D
Ethylene oxide	B(C)**	Propylene oxide	B(C)**
Ethyl ether	C	Styrene	D
Ethyl mercaptan	C	Toluene	D
Formaldehyde (gas)	B	Turpentine	D
Formic acid (90%)	D	Vinyl acetate	D

**For more complete information, see NFPA 497, Recommended Practice for the Classification of Flammable Liquids, Gases, or Vapors and of Hazardous (Classified) Locations for Electrical Installations in Chemical Process Areas, National Fire Protection Association, Quincy, MA, 1997.*

***Group C equipment may be used if all conduits ½ in. in trade size and larger are sealed.*

Table A-5 Ignition Temperature of Common Materials*

	Ignition Temperature			Ignition Temperature	
Material	**°F**	**°C**	**Material**	**°F**	**°C**
Acetaldehyde	347	175	Fuel Oil No. 4	505	236
Acetic acid	867	464	Gasoline	536–880	280–471
Acetone	869	465	Heptane	399	204
Acetylene	581	305	Hexane	437	225
Acrolein (inhib.)	455	235	Hydrazine	75–518	23–270
Allyl alcohol	713	378	Hydrogen	968	520
Ammonia	928	498	Hydrogen sulfide	500	260
Benzene	928	498	Jet fuel (JP-4)	464	240
Butane	550	288	Jet fuel (JP-5)	475	246
Butylamine	594	312	Jet fuel (JP-6)	446	230
Butylene	725	385	Kerosene	410	210
Carbon disulfide	194	90	Methane	999	630
Carbon monoxide	1128	609	Methanol	725	385
Diethylamine	594	312	Methyl acetate	850	454
Ethane	882	472	Methyl acrylate	875	468
Ethanol	685	363	Methyl ethyl ketone	759	404
Ethyl acetate	800	427	Methyl isobutyl ketone	840	440
Ethylamine	725	385	Methyl methacrylate	792	422
Ethylbenzene	810	432	Naphtha (petroleum)	550	288
Ethyl chloride	966	519	Nitromethane	785	418
Ethylene	842	450	Octane	403	206
Ethylene glycol monoethyl ether	455	235	Pentane	470	243
Ethylene oxide	804	429	1-Pentanol	572	300
Ethyl ether	320	160	Propane	842	450
Ethyl mercaptan	572	300	Propylene	851	455
Formaldehyde (gas)	795	429	Propylene oxide	840	449
Formic acid (90%)	813	434	Styrene	914	490
Fuel Oil No. 1	410	210	Toluene	896	480
Fuel Oil No. 2	494	257	Turpentine	488	253
			Vinyl acetate	756	402

**Sources: NFPA 325, Guide to Fire Hazard Properties of Flammable Liquids, Gases, and Volatile Solids, 1994, or NFPA 497, Recommended Practice for the Classification of Flammable Liquids, Gases, or Vapors and of Hazardous (Classified) Locations for Electrical Installations in Chemical Process Areas, 1997, National Fire Protection Association, Quincy, MA, whichever is lower.*

Table A-6 Ignition Temperature for Some Dusts*

Material	Minimum Ignition Temperature	
	°F	°C
Metal Dusts		
Aluminum, atomized collector fines	1022	550
Chrominum (97%), electrolytic, milled	752	400
Ferromanganese, medium carbon	554	290
Iron (98%), H_2 reduced	554	290
Tin (96%), atomized (2% Pb)	806	330
Titanium (99%)	626	330
Vanadium (86.4%)	914	490
Zirconium hydride (93.6% Zr, 2.1% H_2)	518	270
Carbonaceous Dusts		
Charcoal	356	180
Coal, Kentucky bituminous	356	180
Coal, Pittsburgh experimental	338	170
Pitch, coal tar	1310	710
Agricultural Dusts		
Alfalfa meal	392	200
Almond shell	392	200
Apricot pit	446	230
Cinnamon	446	230
Cocoa bean shell	698	370
Coconut shell	428	220
Corn	482	250
Cornstarch, commercial	626	330
Cottonseed meal	392	200
Flax shive	446	230
Gum, arabic	500	260
Hemp hurd	428	220
Lycopodium	590	310
Malt barley	482	250
Milk, skimmed	392	200
Peanut hull	410	210
Pectin	392	200
Potato starch dextrinated	824	440
Rice	428	220
Safflower meal	410	210
Soy flour	374	190
Sugar, powdered	698	370
Wheat	428	220
Wheat flour	680	360
Wood flour	500	260
Yeast, torula	500	260
Chemicals		
Acetoacetanilide	824	440
Adipic acid	1022	550
Benzoic acid	824	440
Dicyclopentadiene dioxide	784	420
Hydroxyethyl cellulose	770	410
Nitrosoamine	518	270
Sorbic acid	860	460
Stearic acid, aluminum salt	572	300
Sulfur	428	220
Drugs		
Aspirin	1220	660
Vitamin B1, mononitrate	680	660
Vitamin C (ascorbic acid)	536	280
Dyes, Pigments, Intermediates		
Green base harmon dye	347	175
Red dye intermediate	347	175
Pesticides		
Benzethonium chloride	716	380
Bis (2-Hydroxy-5-chlorophenyl) methane	1058	570
Crag No. 974	590	310
Ferbam	302	150
Sevin	284	140
Thermoplastic Resins and Molding Compounds		
Acetil Resins		
Acetal, linear (polyformaldehyde)	824	440
Acrylic Resins		
Acrylamide polymer	464	240
Acrylonitrile polymer	860	460

Table A-6 ***(continued)***

Material	Minimum Ignition Temperature	
	°F	°C
Acrylonitrile — vinyl pyridine copolymer	464	240
Methyl methacrylate — polymer	824	440
Methyl methacrylate — ethyl acrylate copolymer	896	480
Methyl methacrylate — ethyl acrylate styrene copolymer	824	440
Cellulosic Resins		
Cellulose acetate	644	340
Cellulose acetate butyrate	698	370
Ethyl cellulose	608	320
Polyethylene resins		
Polyethylene, high-pressure process	716	380
Polyethylene, low-pressure process	788	420
Polyethylene wax	752	400
Styrene Resins		
Polystyrene molding compound	1040	560
Polystyrene latex	932	500
Styrene-acrylonitrile (70–30)	932	500
Vinyl Resins		
Polyvinyl acetate	1022	550
Polyvinyl acetate/ alcohol	1022	550
Polyvinyl butyral	734	390
Vinyl toluene-acrylonitrile butadiene copolymer	936	530
Thermosetting Resins and Molding Compounds		
Allyl Resins		
Allyl alcohol derivative (CR-39)	932	530
Epoxy Resins		
Epoxy	1004	540
Epoxy — bisphenol A	950	510
Phenol furfural	590	310
Phenolic Resins		
Phenol formaldehyde	1076	580
Phenol formaldehyde molding compound (wood flour filler)	932	500
Phenol formaldehyde, polyalkylene-polyamine modified	554	290
Polyester Resins		
Polyethylene terephthalate	932	500
Styrene modified polyester-glass fiber mixture	680	360
Polyurethane Resins		
Polyurethane foam, fire retardant	734	390
Polyurethane foam, not fire retardant	824	440
Special Resins and Molding Compounds		
Cashew oil, phenolic, hard	356	180
Lignin, hydrolized, wood-type, fines	842	450
Petrin acrylate monomer	428	220
Petroleum resin (blown asphalt)	932	500
Rubber, crude, hard	662	350
Rubber, synthetic, hard (33% S)	608	320
Shellac	752	400
Styrene — maleic anhydride copolymer	878	470

**Source: NFPA 499, Recommended Practice for the Classification of Combustible Dusts and of Hazardous (Classified) Locations for Electrical Installations in Chemical Process Areas, National Fire Protection Association, Quincy, MA, 1997.*

APPENDIX B

A Reference for Key Terms

The following list is provided as a tool for the reader to check in order to develop knowledge of significant terms used in the analysis of hazardous locations. This list is not, however, all-inclusive, and the reader is advised to read entire passages of text where terms of interest are referenced.

Index

C

E

F

G

H

I

J

K

L

M

P

Q

R

S

T

U

V

W

Z